EEA Report | No 6/2005

Agriculture and environment in EU-15 — the IRENA indicator report

European Environment Agency

Cover design: EEA
Cover photo © Pawel Kazmierczyk 2005
Layout: Brandpunkt A/S, EEA

This is a joint report by DG Agriculture and Rural Development, DG Environment, Eurostat, DG Joint Research Centre, and the European Environment Agency.
The preparation and production of this report was supported by a grant from the European Commission to the European Environment Agency for the implementation of the IRENA operation.

Legal notice
The contents of this publication do not necessarily reflect the official opinions of the European Commission or other institutions of the European Communities. Neither the European Environment Agency nor any person or company acting on behalf of the Agency is responsible for the use that may be made of the information contained in this report.

All rights reserved
No part of this publication may be reproduced in any form or by any means electronic or mechanical, including photocopying, recording or by any information storage retrieval system, without the permission in writing from the copyright holder. For rights of translation or reproduction please contact EEA project manager Ove Caspersen (address information below).

Information about the European Union is available on the Internet. It can be accessed through the Europa server (http://europa.eu.int).
Cataloguing data can be found at the end of this publication.

Luxembourg: Office for Official Publications of the European Communities, 2005

ISBN 92-9167-779-5
ISSN 1725-9177

© EEA, Copenhagen, 2005

Environmental production
This publication is printed according to high environmental standards.

Printed by Scanprint a/s
— Environment Certificate: ISO 14001
— Quality Certificate: ISO 9001: 2000
— EMAS registered — licence no. DK- S-000015
— Approved for printing with the Nordic Swan environmental label, licence no. 541 055

Paper
— Woodfree matt fine paper, TCF
— The Nordic Swan label

Printed in Denmark

European Environment Agency
Kongens Nytorv 6
1050 Copenhagen K
Denmark
Tel.: +45 33 36 71 00
Fax: +45 33 36 71 99
Web: www.eea.eu.int
Enquiries: www.eea.eu.int/enquiries

Contents

Acknowledgements 5
Executive summary 6
    Background and purpose 6
    General trends in EU-15 agriculture 6
    Agricultural water use 7
    Agricultural input use and the state of water quality 7
    Agricultural land use, farm management (practices) and soils 8
    Climate change and air quality 8
    Biodiversity and landscape 8
    Evaluating agri-environmental indicators and supporting data sets in the EU-15 9

1 Introduction to the IRENA operation 11
    1.1 Policy context and approach 11
    1.2 An indicator framework for agriculture 11
    1.3 Scope and outline 14

2 Agri-environmental indicators 15
    2.1 Introduction 15
    2.2 Indicator development 15
    2.3 Indicator evaluation 24

3 General trends in EU-15 agriculture 26
    3.1 Summary of general trends in EU-15 agriculture 26
    3.2 Introduction 27
    3.3 Trends in cropping and livestock patterns 27
    3.4 Trends in the intensity of farming 30
    3.5 Trends in specialisation and diversification 37
    3.6 Trends in marginalisation and land use change 38
    3.7 Trends in organic farming 40
    3.8 Conclusions: evaluation of indicators 41

4 Agricultural water use 45
    4.1 Summary of main points 45
    4.2 Introduction 45
    4.3 IRENA indicators related to water resources 46
    4.4 Agricultural driving forces 47
    4.5 Agricultural pressures on water resources 47
    4.6 State of/impacts on water resources 48
    4.7 Responses 49
    4.8 Conclusions: evaluation of indicators 50

5 Agricultural input use and the state of water quality 53
    5.1 Summary of main points 53
    5.2 Introduction 53
    5.3 IRENA indicators related to agricultural input use and water quality 55
    5.4 Agricultural driving forces 55
    5.5 Agricultural pressures on water quality 55
    5.6 State of/impacts on water quality 58
    5.7 Responses 61
    5.8 Conclusions: evaluation of indicators 63

Contents

6 Agricultural land use, farm management (practices) and soils **67**
  6.1 Summary of main points 67
  6.2 Introduction 67
  6.3 IRENA indicators related to agricultural land use, farm management and soils 67
  6.4 Agricultural driving forces 68
  6.5 Agricultural pressures on soil 69
  6.6 State of soil 71
  6.7 Responses 74
  6.8 Conclusions: evaluation of indicators 75

7 Climate change and air quality **79**
  7.1 Summary of main points 79
  7.2 Introduction 79
  7.3 IRENA indicators linked to climate change and air quality 80
  7.4 Agricultural pressures on climate change and air quality 81
  7.5 Impact on climate change and air quality 84
  7.6 Responses 84
  7.7 Conclusions: evaluation of indicators 86

8 Biodiversity and landscape **89**
  8.1 Summary of main points 89
  8.2 Introduction 89
  8.3 IRENA indicators related to biodiversity and landscape 91
  8.4 Trends derived from driving force indicators 91
  8.5 Agricultural pressures and benefits on biodiversity and landscapes 91
  8.6 State of/impacts on biodiversity and landscape 93
  8.7 Responses 96
  8.8 Conclusions: evaluation of indicators 101

9 Evaluating agri-environmental indicators and supporting data sets in the EU-15 .... **105**
  9.1 Introduction 105
  9.2 Developing and evaluating agri-environmental indicators 105
  9.3 Review of data sets 107
  9.4 Conclusions 115

List of acronyms **118**

References **120**
  Legislation referred to in the text 122

Annexes **123**
  Annex 1 123
  Annex 2 124
  Annex 3 125

Acknowledgements

The report was prepared by EEA project managers Jan-Erik Petersen and Paul Campling. Peder Gabrielsen provided essential data support and analysis.

The project managers gratefully acknowledge the support from David Hickie and Christina Jacobsen in the writing and editing of this publication.

The IRENA steering group provided helpful guidance and important input throughout the preparation of the report. The steering group members were: Antonio De Angelis, Maria Fuentes Merino, Notis Lebessis, Eric Willems and Nicolas Dandois of DG Agriculture and Rural Development; Bernhard Berger of DG Environment; Koen Duchateau, Derek Peare, Pierre Nadin and Ulrich Eidmann of Eurostat and Jean-Michel Terres and Javier Gallego Pinilla of DG JRC.

The project managers also acknowledge valuable assistance from:

Faycal Bouraoui and Luca Montanarella of DG JRC.

Berien Elbersen, Alterra; Erling Andersen, Danish Centre for Forest, Landscape and Planning, KVL; Frans Godeschalk and Floor Brouwer, LEI-DLO Agricultural Economics Research Institute; Robert Bakker, Agrotechnology & Food Innovations; Nicolas Lampkin, Organic Centre Wales, Institute of Rural Sciences, University of Wales Aberystwyth.

PAIS project team: Ulrike Eppler, Hans-Peter Piorr, Gerd Eiden.

European Topic Centres on Air and Climate Change (ETC/ACC), Biological Diversity (ETC/BD), Terrestrial Environment (ETC/TE) and Water (ETC/WTR).

Lastly, special thanks go to Maria Fuentes Merino of DG Agriculture and Rural Development.

Executive summary

Background and purpose

The IRENA operation (Indicator Reporting on the Integration of Environmental Concerns into Agriculture Policy) is a joint exercise between several Commission directorates-generals (DG Agriculture and Rural Development, DG Environment, Eurostat and DG Joint Research Centre, and the European Environment Agency (EEA) to develop agri-environmental indicators for monitoring the integration of environmental concerns into the common agricultural policy (CAP) in the European Union (EU-15). It is a response of the European Commission to the request of the Agricultural Council to develop a set of indicators for monitoring environmental integration in the CAP.

This report provides an assessment of the progress made in the development and interpretation of the agri-environmental indicators identified in COM (2000) 20 during the IRENA operation. The report builds on more than 35 detailed indicator fact sheets that can be found on the IRENA website: http://webpubs.eea.eu.int/content/irena/index.htm.

All indicators are evaluated according to their usefulness, focusing on key aspects identified by COM (2001) 144: policy relevance, responsiveness, analytical soundness, data availability and measurability, ease of interpretation, and cost effectiveness. A scoring scheme helps to classify the indicators in three categories: 'useful', 'potentially useful' and 'Low potential'. These scores and a more detailed analysis of the strengths and weaknesses of each indicator are compiled in indicator evaluation sheets.

The report also analyses agri-environmental relationships in the EU-15 within the DPSIR framework on the basis of the indicators developed and points to necessary future work in the development of agri-environmental indicators.

In addition, the IRENA operation includes an indicator-based assessment report on the integration of environmental concerns into the CAP. The assessment report builds on the analysis presented in this report and reviews possibilities for, and progress with, environmental integration in EU agriculture policy.

General trends in EU-15 agriculture

The utilised agricultural area (UAA) for EU-12 ([1]) decreased by 2.5 % between 1990 and 2000, affecting mainly permanent grasslands and permanent crops. The total number of livestock units was quite stable from 1990 to 2000 (EU-12), but trends vary for different livestock types and regions. In 1990, 44 % of the agricultural area of EU-12 was managed by high-input farms, but this decreased to 37 % in 2000. Low-input farms occupied the lowest share of the agricultural area (26 %) but this share increased to 28 % in 2000. In some regions the livestock stocking density has increased by more than 10 % mainly due to higher pig stocking density in Denmark, northern Germany, and north-eastern Spain.

Mineral fertiliser use declined from 1990–2001: total nitrogen (N) fertiliser consumption in EU-15 decreased by 12 % and total phosphate (P_2O_5) fertiliser consumption in EU-15 decreased by 35 %. At the same time, the total estimated amount of pesticides used in agriculture increased by 20 % between 1992 and 1999 according to industry figures (ECPA).

Analysis of changes in farm types shows that between 1990 and 2000, the share of the agricultural area in the EU-12 managed by specialised farms increased by 4 %, whereas the area managed by non-specialised farms decreased by 18 %. The largest percentage change is the 'non-specialised livestock' farms type, which has fallen by 25 %. The area under organic farming reached 3.7 % of the total UAA of EU-15 in 2002, up from only 1.8 % in 1998. Organic production accounted for 2 % of EU-15 total production of milk and beef in 2001, but less than 1 % of total production of cereals and potatoes.

Corine land cover (CLC) 1990 and 2000 data shows that the change in land use from agriculture

---

[1] Belgium, Denmark, France, Germany, Greece, Ireland, Italy, Luxemburg, Portugal, Spain, the Netherlands, the United Kingdom.

Executive summary

to artificial surfaces ranges from 2.9 % in the Netherlands to 0.3 % in France. In general the highest percentage of agricultural land (in 1990) converted to artificial areas (in 2000) is found close to major cities.

Indicator evaluation: Five out of thirteen of the indicators used to show agricultural trends are classed as 'useful', while the rest is ranked as 'potentially useful'. In general, the indicators based on FSS, FADN and CLC scored the highest, because these sources provide harmonised regional information. However, it is difficult to link indicators reported at different scales; for example, national data on mineral fertiliser consumption (IRENA 8) with regional data on cropping and livestock patterns (IRENA 13).

Agricultural water use

The irrigable area in EU-12 increased from 12.3 million ha to 13.8 million ha between 1990 and 2000, i.e. by 12 %. In France, Greece and Spain, the irrigable area increased by 29 % during the same period. During the 1990s the reported water allocation rates per ha decreased across the EU-15.

The demand for irrigation water shows a strong regional distribution. From a total of 332 regions, the 41 regions with the highest use of water for agricultural purposes (more than 500 million m^3/year) are all located in southern Europe. The limited data available indicate that the share of agriculture in water use remained stable in the period 1991–1997 in both northern and southern Europe, at about 7 % and 50 %, respectively.

Indicator evaluation: Six out of seven indicators have been evaluated as 'potentially useful'. In spite of a high score for several criteria, the 'water use' indicator (IRENA 10) is classed in the category 'potentially useful' because the trends in the irrigable area are only a proxy indicator for water use intensity. Better data on trends in ground water levels would be very useful, but EU-level data are not available. The indicator (IRENA 31) is therefore classified as having 'Low potential'. Pressure, State/impact and response indicators are underpinned by low or medium quality data and there are weak links between the indicators. Considerable effort is required to improve the indicators throughout the DPSIR framework to increase possibilities for monitoring the impact of agriculture on water resources. Modelling may have a role to play whereby climatic information is combined with crop and land use data to determine water requirements from agriculture.

Agricultural input use and the state of water quality

Nutrient surpluses, as measured by gross nitrogen balance, have generally decreased in the EU-15 between 1990 and 2000. Current nutrient surpluses range from 37 kg N/ha in Italy to 226 kg N/ha in the Netherlands, with four EU Member States showing surpluses above 100 kg N/ha/year (Netherlands, Belgium, Luxembourg and Germany). Since 1990, nitrogen surpluses increased in Spain (47 %) and Ireland (22 %).

National balances can, however, mask important regional differences in the gross nitrogen balance that determine actual nitrogen leaching risk at regional or local level. Individual Member States can thus have overall acceptable gross nitrogen balances at national level but still experience significant nitrogen leaching in certain regions, for example in areas with high livestock concentrations. The calculation of regional of gross nitrogen balances would therefore provide a better insight into the actual likelihood of nutrient losses to water bodies. Such an indicator could not be developed in the timeframe of the IRENA project, mainly due to the lack of important data at regional and national level.

Livestock densities at NUTS 2/3 level give a regionalised picture of likely agricultural nutrient pressure. Regional concentrations of livestock linked to intensive pig and dairy production are found in the west of Germany, the Netherlands, Belgium, Brittany, northwest and northeast Spain, the Italian Po valley, Denmark, the west of United Kingdom and southern Ireland.

Concentrations of nitrates in groundwater have remained largely stable between 1993 and 2002, apart from a significant decline in southern European Member States. Nitrate concentrations at river stations declined slightly between 1992 and 2001 in Denmark, Germany, Luxembourg and the United Kingdom but remained stable in the other four Member States for which data is available. The weighted average share of agriculture in total nitrogen leaching to surface waters for EU-9 (Austria, Belgium, Denmark, Finland, France, Germany, the Netherlands, Italy and Sweden) is 56 %.

Data sets for the indicators pesticide soil contamination (IRENA 20) and pesticides in water (IRENA 30) rely on a modelling approach and case study material, respectively. Key policy responses to agricultural nutrient leaching include

Executive summary

the introduction of codes of good farming practice (GFP) and agri-environment schemes. These can assist in reaching and maintaining the water quality status required under EU-legislation, such as the nitrates directive.

Indicator evaluation: The three indicators classed as 'useful' are: 'area under organic farming' (IRENA 7), 'mineral fertiliser consumption' (IRENA 8) and 'cropping/livestock patterns' (IRENA 13). The other indicators are classed in the category 'potentially useful', including 'gross nitrogen balance', which is not available at regional level. In most cases these indicators have not reached a level of development to be considered as 'useful', because data availability and measurability, and analytical soundness are inadequate. Information on the use and impact of pesticides is in particular difficult to obtain. None of the indicators are, however, regarded as low potential.

Agricultural land use, farm management (practices) and soils

Estimates based on the Pesera model indicate that the areas with the highest risk of soil erosion by water (i.e. more than 5 tonnes soil loss/ha/year) are located in southern and western Spain, northern Portugal, southern Greece and central Italy. Areas with low organic carbon content (0–1 %) appear mostly in southern Europe and correspond to areas with high soil erosion risk. No trend information is currently available. At present the modelling estimates of soil erosion risk and soil organic carbon content have not been updated with information from CLC 2000.

In 2000, approximately 56 % of the EU-15 arable land was covered 70 % of the year and 24 % of the arable land was covered 80 % of the year. Only 4 % and 5 % of the arable area were covered just 40 % and 50 % of the time throughout the year, respectively.

Analysis of land cover changes between CLC 1990 and 2000 indicates that Spain had large land cover flows from forest/semi-natural land to agriculture and from agriculture to forest/semi-natural land. Italy and Portugal showed land cover flow only from agriculture to forest/semi-natural land.

Indicator evaluation: Four indicators in this environmental storyline are classed in the category 'useful': the driving force indicators 'land use change' (IRENA 12), and 'cropping/livestock patterns' (IRENA 13), the pressure indicators 'land cover change' (IRENA 24), and the response indicator 'area under organic farming' (IRENA 7).

Land cover and cropping patterns are ranked the highest. The rest of the indicators are classed in the category 'potentially useful'.

Most of the pressure and all the State/impact indicators have not reached a level of development to be considered as 'useful', mainly due to weaknesses in data availability and measurability as well as analytical soundness. Several of them are obtained via modelling, and efforts are recommended to improve those models to achieve higher robustness and acceptability. To ensure comparable quality between all indicators the State/impact indicators would have to be improved considerably. 'Farm management practices' (tillage methods) (IRENA 14.1) has the lowest score. The information about tillage practices is highly relevant to soil conservation, but little reliable information is available.

Climate change and air quality

While agriculture contributed around 10 % of total greenhouse gas emissions in EU-15 in 2002 it can also function as a sink for CO_2. The main greenhouse gases emitted by agriculture are nitrous oxide and methane, both of which have a far greater global warming potential than carbon dioxide. Agriculture also consumes fossil fuels for farm operations, thus emitting carbon dioxide. Emissions of greenhouse gases by the agriculture sector — methane and nitrous oxide — have fallen by 8.7 % between 1990 and 2002. Within the EU-15, emissions of ammonia from agriculture have also decreased by 9 % between 1990 and 2002 but the sector still provides more than 90 % of total ammonia emissions. In 2003 agriculture contributed 3.6 % of total renewable energy produced and 0.3 % of total primary energy produced in the EU-15.

Indicator evaluation: Most of the indicators (six of the nine) are classed as 'useful'. The response indicators (regional levels of environmental targets and production of renewable energy) are considered as 'potentially useful'. To become useful, their measurability needs to be improved. The energy use (IRENA 11) indicator would be potentially more important if CO_2 emissions would be estimated on the basis of coefficients. The high average score of this storyline arises probably from the national reporting level of pressure and state indicators, and because of the clear targets linked to pressure/State/impact indicators.

Biodiversity and landscape

Extensive farming systems are important for maintaining the biological and landscape diversity of farmland, including Natura 2000 sites. Such systems have been threatened, however, by two different trends: intensification and abandonment. While intensification — in terms of the use of external inputs — seems to have levelled off during the 1990s, the trend towards farm specialisation continues in the EU-15. The decline in the proportion of 'mixed livestock' farms by about 25 % from 1990 to 2000 is particularly significant since these farms are often associated with high biodiversity and landscape quality and form part of high nature value (HNV) farmland. HNV farmland areas are mainly found in Mediterranean regions, upland areas in the United Kingdom and Ireland, mountain areas and parts of Scandinavia, and are estimated to cover around 15–25 % of total UAA in the EU-15.

According to current estimates about 17 % of the habitats in proposed Natura 2000 areas depend on a continuation of extensive agricultural practices. Such management favouring maintenance of biodiversity can be supported via agri-environment schemes and other agricultural policy instruments.

The majority of farmland birds have suffered a strong decline from 1980 to 2002. This decline has levelled off in the 1990s but species diversity remains at a very low level in intensively farmed areas. Data for important bird areas and prime butterfly areas show that a significant share of these sites is negatively affected by agricultural intensification and/or abandonment.

The diversity of agricultural landscapes is difficult to capture in indicators on the basis of current available information. Selected case studies describe typical landscapes (such as bocage and montados). There is great variation between different landscapes in terms of the distribution of arable, grassland, permanent crops and other agricultural uses. Hedgerows and other linear elements are an important feature of most landscapes and were still declining in some regions of the United Kingdom during the 1990s.

Indicator evaluation: Half (eight out of sixteen) indicators are classed as 'useful'. These are the driving force indicators: 'land use change' (IRENA 12), 'intensification/extensification' (IRENA 15), 'specialisation/diversification' (IRENA 16), the pressure indicator 'cropping/livestock patterns' (IRENA 13) and 'land cover change' (IRENA 24), the state indicator on 'populations of farmland birds' (IRENA 28), and the response indicators 'area under nature protection' (IRENA 4) and 'area under organic farming' (IRENA 7).

The indicators 'marginalisation' (IRENA 17), 'genetic diversity' (IRENA 25), 'high nature value (farmland) areas' (IRENA 26), 'landscape state' (IRENA 32), 'impacts on habitats and biodiversity' (IRENA 33), 'impact on landscape diversity' (IRENA 35), 'area under agri-environment support' (IRENA 1) and 'regional levels of good farming practice' (IRENA 2), are considered as 'potentially useful'. The state and impact indicators are weaker than the others because they score lower on the availability of regional and time series data.

Evaluating agri-environmental indicators and supporting data sets in the EU-15

A quarter of the indicators scored 15 points or more implying that they are considered 'useful'. Eight indicators scored between 8 to 14 points and were classified as 'potentially useful', and only one indicator was considered to be of 'Low potential' (ground water levels — IRENA 31) which means that further development appears difficult in spite of the policy relevance of information on ground water levels. However, many indicators in the highest category show deficiencies in some key criteria, mainly due to a lack of time series data or spatial information.

It has also proven difficult to link indicators from different data sets, usually because the reporting levels are not consistent. Data sets in the agricultural domain provide full geographic coverage, time series information and generally high reliability. Thus, farm trend and pressure indicators related to agricultural activity achieve mostly a high score. The existing environmental data sets in the water, soil (and biodiversity) domains are far less developed in terms of coverage, time series and reliability. Consequently, the data required for State/impact indicators are often unavailable. Hence, several indicators of these domains are developed on the basis on modelled or proxy data.

Information at NUTS 2/3 level (where available) is generally sufficient for describing agri-environmental patterns at EU-15 level. However, more detailed level of spatial reporting is needed to understand agri-environmental processes or causal links in sufficient detail for targeted policy action, especially for State/impact indicators and policy responses. In some cases, therefore, it may be more appropriate to adopt a modelling framework, especially where indicators rely on modelled data. Modelling frameworks, which employ a sensitivity analysis, can be used to evaluate the importance of input indicators. This may be more revealing than trying to link up different indicators within a

Executive summary

DPSIR framework. However, at a European scale it is difficult to obtain ground data to calibrate and validate estimates, and even the best modelling cannot improve inadequate input data.

There are many challenges ahead in terms of improving data sets, spatial referencing and ensuring the timely delivery of indicators to policy makers. It is important that the current list of indicators is reviewed and, if necessary, amended to meet future analytical and monitoring needs. This includes deciding which reporting scale is strictly necessary at the EU-15 level, especially in light of the current deficiencies in existing data sets highlighted in the report. The need to extend agri-environmental indicators to include new and future EU Member States has to be taken into account in this regard.

Reporting scale is an important determinant of database and indicator development. Data sets for reporting at EU-level can be coarser than those for national or regional analysis. However, EU indicator data sets are ideally aggregated from more local, spatial information. Data sets should thus be arranged hierarchically. This will allow detailed analysis of agri-environmental issues which, using EU-level data, can only be identified but not analysed. The IRENA operation has made an important contribution to developing agri-environmental indicators at EU-15 level. The cooperation between EU organisations and Member States has proven to be very fruitful in the development of indicators under the IRENA operation. Possibilities for continuing such joint work in the future should be explored.

1 Introduction to the IRENA operation

1.1 Policy context and approach

The European Councils in Cardiff (June 1998) and Vienna (December 1998) stressed the importance of developing environmental indicators to assess the impact of different economic sectors on the environment — including agriculture — and to monitor progress in integrating environmental concerns into Community policies. The European Council in Helsinki (December 1999) adopted the strategy for integrating the environmental dimension into the CAP, previously endorsed by the Agricultural Council (Agricultural Council, 1999). This strategy included a commitment to develop appropriate agri-environmental indicators to monitor such integration. Following this request the Commission issued two communications: COM (2000) 20, which defines the objectives for monitoring the integration process and identifies a set of agri-environmental indicators, and COM (2001) 144, which identifies concepts and potential data sources.

The IRENA operation (Indicator Reporting on the Integration of Environmental Concerns into Agriculture Policy) was launched to further develop agri-environmental indicators for monitoring the integration of environmental concerns into the common agricultural policy (CAP). It is a joint exercise between DG Agriculture and Rural Development, DG Environment, Eurostat and DG Joint Research Centre, and the European Environment Agency (EEA).

The starting point of work in the IRENA operation was the set of 35 indicators defined in COM (2000) 20 and the assessments made by COM (2001) 144 concerning their state of development and required further work. Building on the indicator work, this report presents agri-environmental relationships using the Driving force — Pressure — State — Impact — Response (DPSIR) model. Interlinkages are shown in agri-environmental storylines in relation to major agri-environmental themes: water, land use and soil, climate change and air quality, and biodiversity and landscape. The selected storylines are coherent with the specific objectives of the Agricultural Council's strategy on environmental integration and sustainable development in the CAP.

The report also provides an assessment of the progress made in the development and compilation of the agri-environmental indicators identified in COM (2000) 20 during the IRENA operation (from 13 September 2002 to 28 February 2005). The evaluation of individual indicators focuses on key aspects identified by COM (2001) 144: policy-relevance, responsiveness, analytical soundness, data availability and measurability, ease of interpretation and cost effectiveness. The aim is to assess the suitability of the 35 indicators for monitoring agri-environmental trends as a basis for making policy decisions.

The final chapter summarises the main findings regarding indicator development and evaluation. It assesses the suitability of the data sets used and concludes with recommendations for future work in developing an agri-environmental indicator system.

The report builds on more than 35 [2] detailed indicator fact sheets that can be found on the IRENA website: http://webpubs.eea.eu.int/content/irena/index.htm.

In addition to the present indicator report, the individual fact sheets for over 35 indicators and the underlying databases, the IRENA operation has produced an indicator-based assessment report on the integration of environmental concerns into the CAP. The assessment report — which employs the IRENA indicators — summarises the analysis presented in this report in the context of agri-environmental policy targets. Against that background it then reviews possibilities for, and progress with, environmental integration in EU agricultural policy.

1.2 An indicator framework for agriculture

Section 1.1 explained the demand from the European Council for the Commission to report

---

[2] COM (2000) 20 identifies 35 indicators but five of these are divided into two or three sub-indicators. Moreover, an indicator on the atmospheric emissions of ammonia (IRENA 18b) was added following the request of Member States. Following this approach, the total number of (sub-)indicators reaches 42. These have been evaluated separately.

Introduction to the IRENA operation

on the integration of environmental concerns into Community sectoral policies. The Commission Communication (2000) 20 outlined the type of indicators needed for assessing the integration of environmental concerns into the CAP. An indicator framework for monitoring and evaluating the Agricultural Council Integration Strategy on the basis of DPSIR indicators was proposed (Figure1.1). On this basis, the Communication identified a preliminary set of 35 indicators. The document recognised that there are large gaps in the definition and development of certain indicators — in particular in the areas of farm management, landscape and biodiversity — and stressed that indicators need to be supported by appropriate and reliable statistical information.

The DPSIR concept is an analytical framework that has been developed at the European Environment Agency (EEA, 1999) in order to describe and understand the inter-linkages between economic activities and the environment. It builds on previous OECD work that divided indicators into P S R domains (OECD, 1993). When the DPSIR framework was being developed, one of the objectives was that it should be capable of providing an integrated environmental analysis. This requires the establishment of inter-linkages between driving forces, pressures and impacts. However, the discovery of associations, or even causal links, between different indicators in the DPSIR framework is often hampered by the lack of high quality data sets to underpin the indicators.

The agricultural DPSIR model is a conceptual model that is meant to capture the key 'factors' involved in the relationships between agriculture and the environment and to reflect the complex chain of causes and effects between these factors. However, it should not be overlooked that, as with other models, the agricultural DPSIR model is a simplification of reality. Many of the relationships between agricultural and environmental systems are not sufficiently understood or are difficult to capture in a simple framework. In addition, there are other social and economic factors, which may determine changes in farming systems and rural areas. Such changes may be independent of the current policy response framework and can also affect the environment significantly (Baldock et al., 2000).

In 2001, a second Communication (COM (2001) 144 final) outlined the statistical information needed to develop agri-environmental indicators. For each of the 35 indicators identified by COM (2000) 20, the new Communication proposed brief definitions, the conceptual basis for the indicator, and recommendations for further development.

Figure 1.1 DPSIR framework for agriculture (from COM (2000) 20 final)

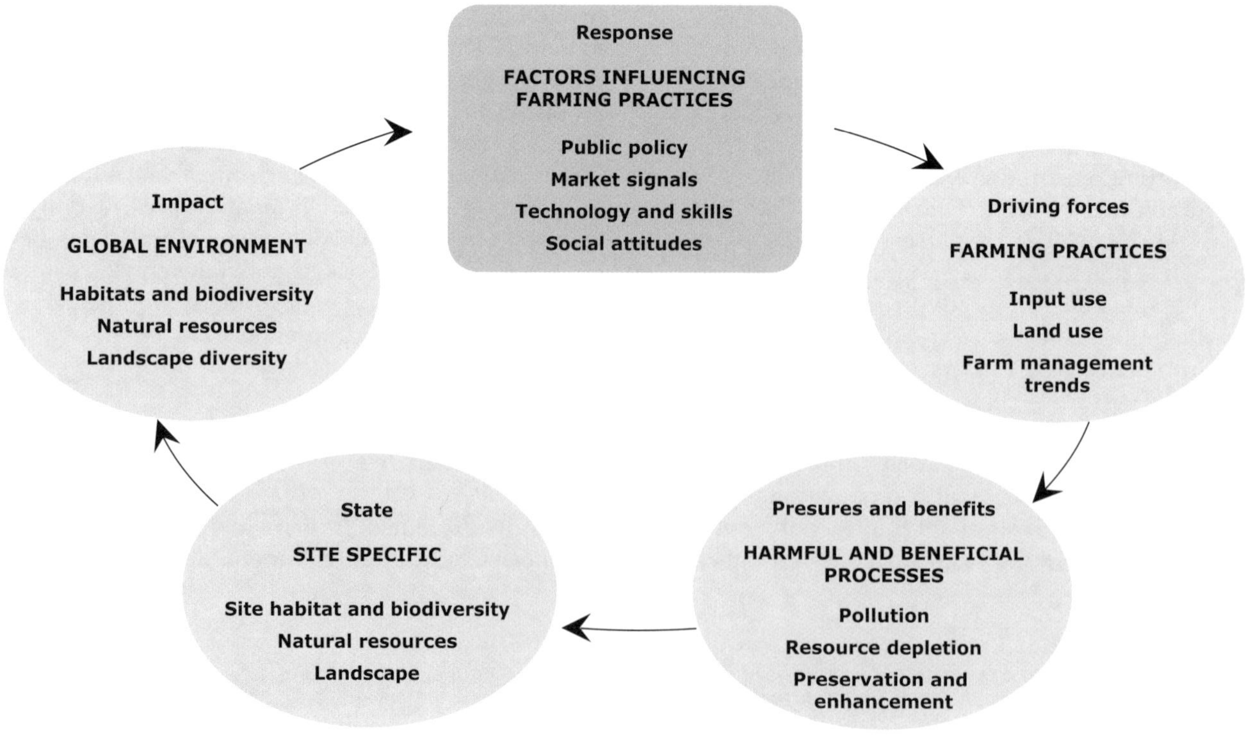

Table 1.1 Explanation of the five domains of the agricultural DPSIR framework and the equivalent IRENA indicators

Domain (³)	Sub-domain	Explanation	No	Indicator
Responses	Public policy	Farming activities are strongly influenced by agricultural and environmental policies and sensitive to input and product price signals. Moreover, changes in technology, farmers' skills, and consumers' and producers' attitudes affect production methods and agricultural practices.	1	Area under agri-environment support
			2	Regional levels of good farming practice
			3	Regional levels of environmental targets
			4	Area under nature protection
	Market signals		5.1	Organic producer prices and market share
			5.2	Organic farm incomes
	Technology and skills		6	Farmers' training levels
	Attitudes		7	Area under organic farming
Driving forces	Input use	A key characteristic of different farming systems and of farming intensity is the use of inputs (fertilisers, pesticides, energy and water).	8	Mineral fertiliser consumption
			9	Consumption of pesticides
			10	Water use (intensity)
			11	Energy use
	Land use	Land use changes as well as cropping and livestock patterns indicate land use intensity and trends in the agricultural sector.	12	Land use change
			13	Cropping/livestock patterns
		Key farm management practices include soil cover, tillage methods and the handling of farm manure.	14	Farm management practices
	Trends	Key trends in farming activities can be expressed at an aggregate level in terms of intensification/ extensification, specialisation/diversification, and economic marginalisation.	15	Intensification/extensification
			16	Specialisation/diversification
			17	Marginalisation
Pressures and benefits	Pollution	Agriculture can lead to nutrient and pesticide residues in soil and water as well as to ammonia and methane emissions. The use of sewage sludge can improve soil fertility but needs to be carefully monitored from a pollution perspective.	18	Gross nitrogen balance
			18sub	Atmospheric emissions of ammonia
			19	Emissions of methane and nitrous oxide
			20	Pesticide soil contamination
			21	Use of sewage sludge
	Resource depletion	Inappropriate use of water and soil leads to environmental pressures. Changes in land cover and genetic diversity can have similar consequences.	22	Water abstraction
			23	Soil erosion
			24	Land cover change
			25	Genetic diversity
	Preservation and enhancement of the environment	Agriculture provides environmental benefits via the management of high nature value farmland and the production of renewable energy.	26	High nature value (farmland) areas
			27	Production of renewable energy (by source)
State	Biodiversity	Birds are a measure of overall species diversity.	28	Population trends of farmland birds
	Natural resources	The state of key natural resources (soil quality, water quantity and quality) needs to be monitored.	29	Soil quality
			30	Nitrates/pesticides in water
			31	Ground water levels
	Landscape	Agriculture has a strong influence on the state of Europe's landscapes through cropping patterns, grazing of upland areas, landscape elements such as hedgerows etc.	32	Landscape state
Impact	Habitats and biodiversity	The share of agriculture in wider environmental issues can be significant. Its impact on natural and landscape diversity is also important but often spatially concentrated and scale-dependent.	33	Impact on habitats and biodiversity
			34.1	Agricultural share of GHG emissions
	Natural resources		34.2	Agricultural share of nitrate contamination
			34.3	Agricultural share of water use
	Landscape diversity		35	Impact on landscape diversity

(³) In several thematic chapters certain indicators are considered to be more usefully employed in a different domain than the one proposed in COM(2000) 20 (e.g. soil erosion as 'state' rather than 'pressure' indicator). This helped to build more 'logical' storylines.

Indicator development in the IRENA operation began with the concepts outlined in COM (2001) 144 final. However, work on indicator definitions and methodology led to the re-naming of some indicators (see Annex 1).

Table 1.1 provides an explanation of the concepts behind the indicators listed under the different domains of the agricultural DPSIR model. In this model, each of the five main domains is split into several sub-domains that are meant to represent the main factors involved. The right column lists the indicators from COM (2001) 144 final with the indicator names adopted in the IRENA operation (Annex 1).

1.3 Scope and outline

This report covers the EU-15 Member States. The reporting target level is administrative regions (Nomenclature of Territorial Units for Statistics — NUTS 2 and 3) across the EU-15, but in some cases data is only available at national level (NUTS 0). To achieve similarly sized regional units, the following NUTS levels were used for the Member States covered:

- NUTS 2: Austria, Belgium, Germany Greece, Luxembourg, the Netherlands, Italy, Portugal and the United Kingdom,
- NUTS 3: Denmark, Finland, France, Ireland, Spain and Sweden.

Where EU-15 wide data sets are not available a case study approach is adopted. Indicators are based on data from a variety of sources and collected at different scales. In general, assessments are based exclusively on IRENA indicator fact sheets, unless other referenced sources are given. Most indicators cover the years between 1990 and 2000. This period includes the MacSharry Reform of the CAP in 1992 as well as the Agenda 2000 CAP reform. Indicators that show trends between 1990 and 2000 are often based on data from the 12 countries that formed the EU in 1990 ([4]) (EU-12).

The report is structured as follows:

- *Chapter 1* — Introduction to the IRENA operation
- *Chapter 2* — Progress in the development of agri-environmental indicators (indicator definitions and purpose; data bases; indicator evaluation criteria)
- *Chapter 3* — Recent farm trends and land use change in EU-15 agriculture
- *Chapter 4* — The pressures of agricultural water use on water resources
- *Chapter 5* — The impact of chemical inputs and organic fertilisers on water quality
- *Chapter 6* — Agricultural land use, farm management and urbanisation, and the state of soils
- *Chapter 7* — Agriculture in the context of climate change, air quality and energy production
- *Chapter 8* — The link of agricultural land use and farm management practices with agricultural landscapes and biodiversity
- *Chapter 9* — Recommendations for improving the data sets supporting agri-environmental indicators.

In the final section of each storyline (Chapters 3 to 8), each indicator is evaluated, based on the scoring scheme presented in Section 2.3.

([4]) Austria, Finland and Sweden joined the EU in 1995.

2 Agri-environmental indicators

2.1 Introduction

OECD (1993) defines indicators as 'parameters, or values derived from parameters, which provide information about the state of a phenomenon/environment/area with significance extending beyond that directly associated with a parameter value'. According to COM (2000) 20, appropriate agri-environment indicators can help to provide information to those involved in the development and implementation of agricultural and rural development policies as well as to the broader public. COM (2000) 20 sets out the following reasons for developing a solid set of agri-environmental indicators:

- to help monitor and assess agri-environmental policies and programmes, and to provide contextual information for rural development in general;
- to identify environmental issues related to agriculture;
- to help target programmes that address agri-environmental issues;
- to understand the linkages between agricultural practices and the environment.

The development of indicators that reflect the above needs and fulfil minimum quality criteria is a demanding task. Section 2.2 explains the approach taken in the IRENA operation to the stepwise development of the 35 indicators and associated fact sheets that underpin the analysis in this report. Under the IRENA operation three types of fact sheets were produced to provide a means of communicating the progress and development to experts at EU level and in the Member States:

- Methodology fact sheets present the indicator definition, concept, limits, potential data sources, and references;
- Methodology/data fact sheets present the data (in the form of tables, figures or maps), methodological approach and preliminary results, including an assessment of the data sources and an exploratory analysis of the results.
- Indicator fact sheets present the data and the indicator assessment (summarised at the beginning of the fact sheet in the form of key messages), including agri-environmental context and policy relevance, data gaps and possibilities for indicator improvement. A meta data section is included in which the data is described and their pertinence and quality is evaluated.

The indicator fact sheets can be found on the IRENA website (http://webpubs.eea.eu.int/content/irena/index.htm).

Indicators have to fulfil certain quality criteria in order to be useful for environmental analysis and to support the policy-making process. The scoring system developed in the IRENA operation on the basis of criteria set out in COM (2001) 144 is explained in Section 2.3.

2.2 Indicator development

The following sections provide information on the level of development of the agri-environmental indicators listed in COM (2001) 144, which are structured along the lines of the indicator domains outlined in COM (2000) 20. Indicators are grouped under the main DPSIR domain and sub-domain. The agri-environmental analysis developed in the following sections shows, however, that this classification is not always very clear-cut as indicators can shift their role within the DPSIR framework depending on the agri-environmental context. One example for such a shift is the placing of the input use indicators in the pressure category in the biodiversity chapter (see Section 8.3). An explanation of the definition, data sets, geographical reporting level and time series used for each indicator is provided in Annex 3.

2.2.1 Driving force indicators

The driving force indicators contribute to a better understanding of the state and evolution of regional farming systems in relation to input use, land use and management practices. They also shed light on general farm trends (intensification/extensification, diversification, and marginalisation) that can affect the conservation of environmental resources in either positive or negative ways.

2.2.1.1 Input use indicators

Input use indicators are developed to monitor the trends in the use of agro-chemicals (IRENA 8

— Mineral fertiliser consumption and IRENA 9 — Consumption of pesticides), trends in the use of water (IRENA 10 — Water use (intensity)) and trends in the use of energy (IRENA 11 — Energy use).

IRENA 8 — Mineral fertiliser consumption
The evolution of the consumption of nitrogenous (N) and phosphate (P) mineral fertilisers over time is based on Faostat data, which provides annual information at the Member State level (NUTS 0) between 1990 and 2002. Change data between 1990 and 2001 is based on rolling averages of three years (average of 1989,1990,1991 volumes and average 2000, 2001, 2002 volumes), to minimise the effect of weather and agronomic conditions. The sub-indicator shows the range of fertiliser application rates for a variety of crops, on the basis of information provided by the European Fertiliser Manufacturers Association (EFMA). This information could be used further to estimate regional fertiliser application rates, but would need to take into account regional farming practices, soil type, and climate.

IRENA 9 — Consumption of pesticides
The consumption of pesticides, which includes all plant protection products apart from biocides and disinfectant products, is indicated by used and sold quantities of different pesticide categories (tonnes of active ingredient, a.i.). The European Crop Protection Association (ECPA) supplies pesticide use data broken down by active ingredient, main crop and Member State. Member States supply pesticide sales data (tonnes of active ingredient) broken down in four use classes (herbicides, fungicides, insecticides and other pesticides). The sub-indicator shows the range of pesticide application rates of different pesticide categories (kilograms of active ingredient per hectare) by dividing volumes by the utilised agricultural area. Annual sales data is provided for the period 1992–2002, and annual use data is provided for 1992–1999.

IRENA 10 — Water use (intensity)
Trends in irrigable area (area covered with irrigation infrastructure) and trends in total area irrigated at least once a year (actual area irrigated) are used as proxy indicators of water use (intensity). The Farm Structure Survey provides information on the irrigable area for all EU-15 Member States, but the information on farms reporting to have irrigated at least once during the year is only available for southern EU-15 Member States. Information is provided for the ten most important crops irrigated (durum wheat, grain maize, potatoes, sugar beet, sunflower, soy, fodder plants, fruit and berries, citrus fruit and vines). Information is reported at NUTS 2 and 3 levels for 2000, and a trend analysis is based on data from 1990, 1993, 1995, 1997 and 2000. The sub-indicator shows the change in total irrigable area as share of UAA. This indicates the importance of irrigated agriculture within a Member State, and how this is evolving with time.

IRENA 11 — Energy use
Direct energy use in primary production is mainly related to heating (e.g. through the use of oil and electricity) and the use of machinery (e.g. transport with tractors). Energy use is indicated by the annual use of energy at farm level by fuel type (GJ/ha), and the energy used to produce mineral fertilisers for agricultural use (GJ/ha). FADN data is used to derive the cost of energy (by fuel type) as a proxy of energy use. Energy costs per ha of utilised agricultural area and per 100 euro of output are calculated for Member States and some regions (NUTS 0 and 1) for 1990 and 2000. SIRENE data provides the actual final energy consumption in agriculture by fuel type at Member States level (NUTS 0). The energy used to produce mineral fertilisers for agricultural use is estimated using industry figures in the Netherlands, which is then extended to the rest of EU-15 using fertiliser use data (IRENA 8).

2.2.1.2 Land use indicators

Land use indicators are developed to monitor the impact of urbanisation on agricultural land (IRENA 12 — Land use change) and the trends in agricultural land use (IRENA 13 — Cropping/livestock patterns).

IRENA 12 — Land use change
Area of land use change from agriculture to artificial surfaces between 1990 and 2000 is derived by using the Corine 1990 and 2000 land cover databases. The Land and Ecosystems Accounts (LEAC) method is used to produce the land use change indicator on the basis of the Corine land cover change database, which is constructed on the basis of 5 ha land cover changes detected from satellite images and ancillary resources (aerial photographs, ground truthing etc.). The sector share of land converted from agriculture to artificial surfaces (%) indicates which sectors are encroaching on agricultural land. Both indicators are reported at regional level (NUTS 2 and 3), but could be reported at a much finer scale if required. At the time of writing, data for the following Member States and regions had been processed: the Netherlands, Ireland, Belgium, France, Italy, Portugal, Denmark, Luxembourg, Spain and Germany. Changes in the United Kingdom, Sweden, Austria, Finland and Greece are not analysed in this report, as data was not available at the time of writing.

IRENA 13 — Cropping/livestock patterns
Cropping patterns are indicated by trends in the share of the utilised agricultural area occupied by the major agricultural land uses (arable, permanent grassland and permanent crops). Livestock patterns are indicated by trends in the share of major livestock types (cattle, sheep and pigs). In both cases indicators are derived from Farm Structure Survey data. The different types of animals are standardised by using coefficients to derive livestock units (Eurostat, 2004). The coefficients take into account the feeding regime for different livestock categories and age. Indicators are reported at the regional level (NUTS 2 and 3) for 1990 and 2000, percentage changes are calculated.

2.2.1.3 Management indicators

There is only one management indicator (IRENA 14 — Farm management practices).

IRENA 14 — Farm management practices
Farm management practices are defined as the decisions and practical operations that shape the practical management of farms. The indicator is developed to include information on soil cover on arable land, tillage systems, and types and capacity of storage facilities for organic fertilisers. Data on soil cover on arable land was provided by the PAIS project, derived from information on the seeding and harvest dates of arable crops in combination with cropping area (cereals, oilseeds, dry pulses, potatoes, sugar beet, tobacco and other industrial crops, forage crops and temporary pastures) based on Farm Structure Survey data (PAIS II, 2005). Information on different tillage systems is also based on information compiled in PAIS II, while the sub-indicator on types and capacity of storage facilities for organic fertilisers is based on FSS data.

2.2.1.4 Trend indicators

Trend indicators are developed to monitor changes in the level of farm intensification or extensification (IRENA 15 — Intensification/extensification) and in the degree of farm specialisation and diversification (IRENA 16 — Specialisation/diversification). In addition, farm economic and demographic data is used to indicate regional changes in farm marginalisation (IRENA 17 — Marginalisation).

Three farm typologies have been developed in the IRENA operation to help characterise general regional trends. These are required to reflect the different dimensions (input use, farm system, specialisation) that need to be explored in a farm trend analysis. The first typology (related to intensification/extensification) differentiates farms according to the expenditure on purchased farm inputs (low, medium and high input farms), using data from FADN. Expenditure is regarded as a proxy for input use.

The second typology differentiates farms based both on the Community Typology of agricultural holdings and land use criteria, using data from FADN to differentiate holdings according to their type of farming (e.g. grazing livestock, cropping — specialist crops, horticulture etc.).

A third typology is used for the specialisation/diversification indicator, which groups the Community Typology farm types into specialised and non-specialised categories. Further detail on the typologies developed is provided in Annex 2.

IRENA 15 — Intensification/extensification
Intensification/extensification is indicated by: trends in the share of agricultural area managed by low-input, medium-input or high-input farm types (based on the average expenditure on inputs per hectare), livestock stocking densities, and trends in yields of milk production and cereals. Data is based on FADN data (NUTS 0 and 1), and trends are derived from differences between 1990 and 2000.

IRENA 16 — Specialisation/diversification
Specialisation is indicated by trends in the share of the agricultural area managed by specialised types of farm. The share of agri-environment payments in gross farm income is developed as a proxy indicator of diversification. Data is based on both FSS and FADN data (NUTS 0 and 1), and trends are derived from differences between 1990 and 2000.

IRENA 17 — Marginalisation
The indicator of marginalisation links economic and demographic factors driving marginalisation and identifies key regions at 'double risk'. Marginalisation is estimated to occur in regions where farming generates low profitability and farmers are close to retiring age. Regions of low profitability are defined as those where more than 40 % holdings have a farm net value added per annual work unit (FNVA/AWU) below 50 % of the average FNVA/AWU of the region. Regions with a high share of farmers close to retiring age are defined by a share of holdings with farmers aged 55 years and over that exceeds 40 %. Data is based on FADN data (NUTS 0 and 1), and trends are derived from differences between 1990 and 2000.

2.2.2 Pressure indicators

The pressure indicators aim at identifying harmful and beneficial processes attributed to agriculture. These are subdivided into three sub-domains: pollution, resource depletion, and benefits.

2.2.2.1 Pollution indicators

Pollution indicators are developed to monitor non-point source levels of nitrogen from agriculture (IRENA 18 — Gross nitrogen balance) (5), air emissions of ammonia from agriculture (IRENA 18sub — Atmospheric emissions of ammonia), air emissions of methane and nitrous oxide from agriculture (IRENA 19 — Emissions of methane and nitrous oxide), the potential annual average content of herbicides in soils (IRENA 20 — Pesticide soil contamination), and the concentrations of heavy metals in sewage sludge spread on agricultural land (IRENA 21 — Use of sewage sludge).

IRENA 18 — Gross nitrogen balance

Gross nutrient balance relates to the potential surplus of nitrogen and phosphorus on agricultural land. This is estimated by calculating the balance between nitrogen and phosphorus (Figure 2.1) added to the agricultural system and nitrogen and phosphorus removed from the system per hectare of agricultural land. The gross nitrogen balance indicator accounts for all inputs and outputs on the soil surface, and includes all residual emissions of nitrogen from agriculture into soil, water and air. The volatilisation of ammonia is therefore included. The principle inputs include volumes of nutrients as inorganic fertiliser, livestock manure, nitrogen fixation by crops and atmospheric deposition per hectare. The principle outputs include volumes of nutrients taken out by harvested crops and grass/fodder eaten by livestock per hectare. The indicator is based on either balances submitted to the OECD or calculated by the IRENA team using EU-15 wide data sets. Given available resources and data limitations the indicator could only be developed at the national level for 1990 and 2000.

A full explanation of the nutrient balances is provided by the OECD/Eurostat Nitrogen and Phosphorus handbooks (see OECD 2006, forthcoming).

Data has been extracted from the spreadsheets provided by Member States to the OECD. The United Kingdom, Ireland, Belgium (Wallonia), Spain, Greece, Denmark and Luxembourg have not provided data. Sweden has provided national and regional balances, with only a breakdown of balances for arable land. France provided national balances, but without including nutrients from atmospheric deposition. There is ongoing national work to estimate nutrient balances.

National balances have been calculated for Member States not providing information to the OECD. The following data sources were used: Harvested crops and forage (Eurostat's ZPA1 data set); Livestock numbers (Eurostat's ZPA1 data set or Farm structure

Figure 2.1 Terms of the gross nitrogen balance

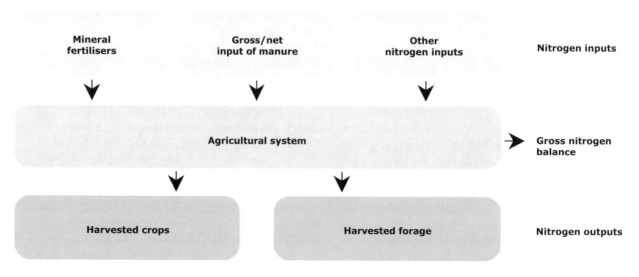

(5) A phosphorus balance indicator could not be developed for technical reasons.

survey); Livestock excretion rates (OECD or averaged coefficients from Member States); Fertiliser rates (EFMA); Nitrogen fixation (OECD or averaged coefficients from Member States); Atmospheric Deposition (EMEP); Yields (Eurostat's ZPA1 data set or average coefficients from Member States).

Coefficients for Spain and Greece are based on the average of coefficients used in Italy and Portugal. Coefficients for Denmark, the United Kingdom, Luxembourg, Belgium (Wallonia and Flanders) and Ireland are based on the average of coefficients used in Germany, the Netherlands and Belgium (Flanders). In addition, a balance was calculated for France that includes N deposition from the atmosphere.

IRENA 18sub — Atmospheric emissions of ammonia
This indicator shows the annual atmospheric emissions of ammonia (NH_3) in the EU-15 for 1990–2002, and the contribution of agriculture to total emissions of ammonia in 2002. The indicator is based on officially reported 2004 national total and sectoral emissions to UNECE/EMEP Convention on Long-range Transboundary Atmospheric Pollution. However, the data is only reported at Member State level (NUTS 0).

IRENA 19 — Emissions of methane and nitrous oxide
Aggregated annual emissions of methane (CH_4) and nitrous oxide (N_2O) from agriculture. Emissions are shown relative to 1990 baseline levels expressed as CO_2 equivalents. The indicator is based on data from the official national total, sectoral emissions, livestock and mineral fertiliser consumption data reported to UNFCCC under the EU monitoring mechanism and Eionet. However, the data is only reported at Member State level (NUTS 0).

IRENA 20 — Pesticide soil contamination
The indicator 'Pesticide soil contamination' uses a model to calculate the potential average annual content of herbicides in soils. Due to data limitations, the five most used herbicides per region are used as a proxy for total pesticide use. However, herbicides are also a pesticide category that is particularly relevant from a soil protection perspective. The figures are calculated assuming an exponential decay of the active ingredients taking into account average monthly temperatures. Potential average soil pesticide content is also affected by the rates at which pesticides are applied. Pesticide application rates are estimated on the basis of Eurostat pesticide statistical data (2002) and Farm Structure Survey data (1997, 2000). The calculated time series (1993–1997) of potential annual average herbicide soil concentrations are analysed to detect statistically significant trends.

IRENA 21 — Use of sewage sludge
The indicator focuses on the use of sewage sludge in agriculture as sufficient monitoring data on heavy metal or organic pollution in water is not available. It relates therefore less to 'water contamination', which was the original concept of COM (144) 2001, than to the recycling of waste in agriculture. However, sewage sludge contains heavy metal concentrations that need to be monitored carefully. The indicator builds on data on volumes and heavy metal concentrations of sewage sludge that are submitted by Member States to the European Commission in the context of the requirements under the Standardised Reporting Directive (91/692/EEC) (1995–2000).

2.2.2.2 Resource depletion

Resource depletion indicators aim to monitor possible pressures on natural resources from agricultural activities, in terms of use of water (IRENA 22 — Water abstraction) and soil management (IRENA 23 — Soil erosion), maintenance of natural and semi-natural habitats (IRENA 24 — Land cover change), and biodiversity (IRENA 25 — Genetic diversity).

IRENA 22 — Water abstraction
Water abstraction by agriculture is indicated by the annual water allocation rates for irrigation (m^3/year/ha). These are derived from the reported water abstraction rates (m^3/year) from the Joint OECD/Eurostat questionnaire and statistics on irrigable area (ha) from the Farm Structure Survey. The indicator is reported at Member State level (NUTS 0) from 1990 to 2000. Estimated regional water abstraction rates for irrigation (m^3/year) are derived by weighting reported national rates according to regional data on irrigable area. This sub-indicator is reported at NUTS 2 and 3 for 2000.

IRENA 23 — Soil erosion
Annual soil erosion risk by water is estimated on the basis of the Pesera model. Pesera is a physically based model that uses the following data as input: Corine land cover (Land use), GTOPO30 (Relief), MARS database (Climate), European Soil Database (Soil). The indicator is reported at NUTS 2 and 3, but no trends are indicated.

IRENA 24 — Land cover change
The area of the entries and exits to and from agricultural and forest/semi-natural land between

1990 and 2000 is derived by using the Corine 1990 and 2000 land cover databases. The Land and Ecosystems Accounts (LEAC) method is used to produce the land use change indicator on the basis of the land cover change database, which is constructed on the basis of 5 ha land cover changes detected from satellite images and ancillary resources (aerial photographs, ground truthing etc.). Net arable and permanent crop and pastureland cover changes between 1990 and 2000 indicate important land cover changes within agriculture. Both indicators are reported at regional level (NUTS 2 and 3) — but could be reported at a much finer scale if required. Data from the following Member States were used: the Netherlands, Ireland, Belgium, France, Italy, Portugal, Denmark, Luxembourg, Spain and Germany. Changes in the United Kingdom, Sweden, Austria, Finland and Greece are not analysed in this report, as trend data were not available or not yet processed at the time of writing.

IRENA 25 — Genetic diversity
Genetic diversity is indicated by the distribution of risk status of national livestock breeds in agriculture. The indicator is based on FAO's Domestic Animal Diversity Information System (DAD-IS), July 2003 update. The reporting level is NUTS 0.

2.2.2.3 Benefit indicators

The maintenance of high nature value farmland (IRENA 26) and the increased production of renewable energy from agricultural sources (IRENA 27) are both regarded as environmentally beneficial. However, the production of energy crops from agriculture can also have negative environmental effects due to possible changes in land use and the level of agricultural inputs applied.

IRENA 26 — High nature value (farmland) areas
High nature value farmland comprises the core areas of biological diversity in agricultural landscapes. They are often characterised by extensive farming practices, associated with a high species and habitat diversity or the presence of species of European conservation concern. This indicator shows the share of the utilised agricultural area that is estimated to be high nature value farmland. The indicator is based on Corine land cover and the farm accountancy data network (FADN). The indicator is reported at NUTS 0, but no trends are indicated.

IRENA 27 — Production of renewable energy (by source)
The production of renewable energy from agricultural sources is indicated by: land use devoted to energy/biomass crops, and primary energy produced from crops and by-products. This is based on a variety of data sources. The indicator is reported for 2003 at NUTS 0 level, but no trends are indicated.

2.2.3 State/impact indicators

The IRENA state and impact indicators are described together in this section. However, their scope and scale of reporting (regional for state, national/EU for impact) are different. The state indicators are meant to describe the state of different natural and semi-natural resources in rural areas. The impact indicators serve to identify the share of agriculture, as a sector, to undesirable changes in the state of the environmental resources (e.g. nitrate contamination), as well as its effective contribution to the preservation/enhancement of other environmental resources (e.g. landscape diversity). Sometimes state and impact domains are very closely related (in particular for biodiversity) and they are presented together in the agri-environmental chapters (in the tables they are separated by a dotted line).

3.2.3.1 Biodiversity

In the IRENA set, there is one state indicator: (IRENA 28) and one impact indicator (IRENA 33) related to biodiversity. However, in practice they reflect the status of two types of animal species closely linked to agricultural areas for which data is available: farmland birds and butterflies.

IRENA 28 — Population trends of farmland birds
The trend is calculated for EU-15 and is based on population data for 23 species of farmland birds characteristic of farmland areas all over Europe. The trends are the result of aggregations on national and regional level weighted by bird population sizes. Moreover trends have been estimated when data was missing using the TRIM programme (Pannekoek and van Strien, 1998). The data originates from national monitoring data collected by the Pan-European Bird Monitoring project. The project is coordinated by the Royal Society for the Protection of Birds (RSPB), the European Bird Census Council (EBCC) and BirdLife International. The information was collected in 18 countries, 11 of which are EU-15 Member States. Time series data from 1990 are available for the following EU-15 countries: Belgium, Denmark, Germany, France, Sweden and the United Kingdom. The following countries joined the survey after 1990: The Netherlands (1991), Spain (1996), Austria and Ireland (1998), and Italy (2000).

Whereas the main indicator shows the trend of the populations of farmland bird species collectively, the sub-indicator shows the share of species, which have declining populations (more than 20 % within the period 1970–2000). The data for the sub-indicator come from national estimates of the overall trends in bird population sizes from 1970–2000 (BirdLife International/EBCC, 2000) and cover the same species of farmland birds as the main indicator.

IRENA 33 — Impacts on habitats and biodiversity

The indicator shows the national proportion of important bird areas (IBAs) (Heath and Evans, 2000) that are reported to be affected by agricultural intensification or abandonment. Data derives from the important bird areas programme of BirdLife International and is compiled in a central database managed by BirdLife International. The organisation uses a standardised questionnaire to collect quantitative and qualitative information from IBA coordinators on land use, habitats and bird species, potential or actual threats and other factors for each IBA. The national IBA data used for the indicator are not fully comparable due to the expert nature of some of the information supplied but are considered to give a good picture at EU-15 level.

The second indicator shows the population trends of agriculture-related butterfly species in prime butterfly areas (PBAs). PBAs have been identified on the basis of a European expert survey organised by Butterfly Conservation International (data reported in Van Swaay and Warren, 2003). The selection of relevant butterfly species builds on their classification in international or European conservation legislation or conventions. Data on trends of species and on threats related to PBAs are based mainly on expert judgment and not on quantitative time series. Interpretations may thus differ from one Member State to another.

2.2.3.2 Natural resources (soil, water)

The state of water in terms of quality and quantity is reflected in indicators IRENA 30 and 31 while IRENA 29 is about soil quality. The share of agricultural activities on climate change emissions, the use of water and nitrate contamination is presented by sub-indicators of IRENA 34.

IRENA 29 — Soil quality

In the absence of an agreed definition of soil quality, organic carbon content (percentage) of the topsoil (0–30 cm) has been defined as the indicator for soil quality. Up to a certain extent, high organic carbon content corresponds to good soil conditions from an agri-environmental point of view: limited soil erosion, high buffering and filtration capacity, rich habitat for soil organisms, enhanced sink for atmospheric carbon dioxide. The data is the result of a model to calculate the degradation of organic carbon content, which uses several data sets: the European Soil Database (soils types, texture), Corine land cover (land cover), and Global Historical Climatology Network (temperatures). The indicator aims to produce baseline data (e.g. a data layer of existing organic carbon content) rather than modelling soil development and carbon stocks.

IRENA 30 — Nitrates/pesticides in water

The indicator shows annual trends in the concentrations of nitrates (mg/l N) in ground and surface water bodies at EU-15 level. The content of pesticides in water is indicated by annual trends in selected pesticide compounds (µg/l). Both data is provided at Member State level for the period 1992–2001 and come from the Eurowaternet network managed by EEA. However, there are major gaps in the monitoring of the status of groundwater, rivers and lakes across the European Union as well as difficulties of comparability because of different sampling methods.

IRENA 31 — Ground water levels

Data on trends of groundwater levels could not be obtained. A case study based in Spain (trends 1980–1998 of ground water levels of one aquifer) has been used in the agri-environmental storyline related to water resources.

IRENA 34.1 — Agricultural share of GHG emissions

The share of the agricultural sector to total EU-15 emissions of the greenhouse gases CO_2, CH_4, and N_2O is obtained from official national sectoral emissions, livestock and mineral fertiliser consumption data reported to UNFCCC under the EU monitoring mechanism and the Eionet network of the EEA (1990 to 2002).

IRENA 34.2 — Agricultural share of nitrate contamination

The share of agriculture in nitrate contamination is reported by some Member States in response to the OECD questionnaire underpinning the forthcoming report on Environmental Indicators for Agriculture Volume 4. The questionnaire requests information on the contamination of surface, ground and coastal waters.

IRENA 34.3 — Agricultural share of water use

Data for some Member States is available through the joint OECD/Eurostat questionnaire (years 1990 and 1998). These were used to compile the change in the proportion of agricultural water use of surface and ground waters compared to other economic sectors (electricity, industry, public water supply). Time series analysis could only be carried out for some EU-15 Member States.

2.2.3.3 Landscape

One state (IRENA 32) and one impact (IRENA 35) indicator related to landscapes were proposed in COM (144) 2001.

IRENA 32 — Landscape state
The landscape state indicator shows the variety of agricultural landscapes across Europe by analysing selected landscape parameters (presence of crops, linear elements, and patch density) with strong links to agricultural land use. These parameters have been calculated for selected regional case study areas representative of different European landscapes. These are for instance: *montados* of Portugal, *open field* landscapes in the central plateau of Spain, *bocage* in France, *highlands* of Scotland. The following data sets have been used to derive the different parameters:

- *FSS: the percentage of agriculture crop types in total land area* shows the contribution of each of the crop types (arable land, grassland and permanent crops) to the total amount of land surface.
- *CLC number of agricultural classes* illustrates the diversity of land cover types in each area.
- *Corine land cover patch density* provides an indication of the fragmentation of agricultural land. This is linked to the diversity of different land cover/uses in a certain area.
- *LUCAS: the number of linear elements* indicates the number of agriculturally linked linear elements per kilometre on the basis of transect observations.

IRENA 35 — Impact on landscape diversity
This indicator presents the evolution of some of the parameters calculated in IRENA 32. The changes of crop type distribution (e.g. arable, grasslands) and patch density are shown for selected landscape types. In addition, the indicator contains data on changes (from 1990 to 1998) in total linear landscape features (km) in England, Wales and Scotland (based on the UK Countryside Survey) and in selected reference sites of Sweden (based on the LiM project).

2.2.4 Response indicators

These indicators are aimed at analysing societal, market and policy responses that influence production systems and agricultural practices. Ideally, the responses reflect information derived from state and impact indicators.

2.2.4.1 Public policy

Some of the main policy measures to address environmental problems in agriculture have been translated by COM (2000) 20 into response indicators related to the public policy dimension.

IRENA 1 — Area under agri-environment support
Agri-environment measures are a compulsory part of EU rural development policy (Council Regulation (EC) 1257/99) and can be considered a core instrument for the integration of environmental goals into the CAP. The indicator measures trends in the agricultural area enrolled in agri-environment measures and its share of utilised agricultural area (UAA) between 1998 and 2002 (last available data). The 2002 data includes all the new contracts signed in 2000, 2001 and 2002 under Regulation 1257/1999 as well as the on-going commitments under the predecessor Regulation 2078/92 which still represent a considerable proportion of the total in some countries. However, for the schemes under the old Regulation 2078/1992, only the total area and the area under organic farming are available. The regional share of agricultural land enrolled in agri-environment measures in total UAA in 2001 is also provided (where possible). A breakdown of total area under agri-environment agreements by main type of action (2002) is included. This indicator provides insight into the environmental policy priorities at national or regional level. The Common Indicators for monitoring the implementation of rural development programmes (RDPs) of 2001 and 2002 (tables f and 7) form the data basis for the indicator.

Sub-indicators show the trends of annual expenditure on agri-environment measures per ha of UAA (2000–2003), based on budgetary data, and the number of endangered breeds under agri-environment measures (2001), based on the Common Indicators for monitoring the implementation of RDPs.

IRENA 2 — Regional levels of good farming practice
This indicator describes the range and type of relevant categories of farming practices covered by the codes of good farming practice (GFP) defined by Member States in their rural development programmes (period 2000–2006). The national/regional codes of GFP included in RDPs as a reference level for applying for agri-environment measures and LFA compensatory allowances have been used. A list of categories of agricultural practices/environmental issues potentially covered by the codes was drawn up as a benchmark table.

The main farming practices considered refer to: soil management, irrigation, fertilisation and plant protection management, waste management, pasture management, biodiversity and landscape. The sub-indicators explore: the 'regulatory' (requirements based on legislation) or 'advisory' approach (based on recommendations) taken by Member States in preparing their code of GFP, and the share of GFP requirements being verifiable standards (which are subject to control).

IRENA 3 — Regional levels of environmental targets

Based on European Commission and national policy documents, the indicator lists the environmental targets set at EU or Member State level which are relevant to agriculture for a range of environmental issues (climate change, air, pesticides, water, organic farming). Originally, the purpose of this indicator was to identify environmental targets for agriculture at regional level and document the relative success of reaching them. However, this approach was not feasible due to lack of regional information and monitoring data. Instead the indicator focuses on the existence of national and EU level targets as well as associated action plans. Therefore, the indicator does not involve an assessment of success in meeting the targets but simply describes in which environmental areas they exist.

IRENA 4 — Area under nature protection

Based on data from sites proposed under the Habitats Directive (referred to as 'Natura 2000 sites'), the indicator shows the proportion of the Natura 2000 sites covered by 'targeted habitats'. The targeted agricultural habitat types are defined as the habitats in the Habitats Directive Annex I that depend on a continuation of extensive farming practices. The process of selecting sites is not yet completed and the analysis is based on data from July 2004.

2.2.4.2 Market signals

Indicator IRENA 5 compares key economic parameters between organic and conventional farms, on the basis that financial viability is a key determinant of both uptake and maintenance of organic farming. The indicator is split into two sub-indicators (5.1 and 5.2).

IRENA 5.1 — Organic producer prices and market share

Organic producer prices and market share indicate levels of consumer demand for organic products and market signals to organic producers. Originally the indicator focused exclusively on organic price premia (the difference between the prices of products produced with conventional and organic agriculture). However, in the development of the indicator the market share of organic products became recognised as a relevant driving factor behind the adoption of organic methods by farmers. Additionally, market share provides a more stable indication of market development and consumer willingness to buy organic products. Premium prices are an important contributor to organic farm incomes, but taken on their own are not necessarily a good indicator of the financial viability of organic farming, or of market conditions, as they may be a result of declining conventional prices rather than increasing organic prices. Both data have been compiled by the EU funded research project (QLK5-2000-01124) organic marketing initiatives and rural development (OMIaRD) for the years 2000 and 2001 and are not derived therefore, from official EU data sets.

IRENA 5.2 — Organic farm incomes

The farm incomes of organic farms compared to similar conventional farms are used to indicate the combined impacts of prices, agri-environment support payments and other factors on the financial viability of organic holdings. The data come from the farm accountancy data network (FADN) of some Member States, which include in this network of agricultural farms a sample of organic farms (identified with a particular code included in the EU farm return since 2000). The data corresponds to the year 2001, in which there is better coverage than in 2000. Trends in organic farm income are based on national FADN data for Austria and Germany. The income parameters chosen are the farm net value added per agricultural working unit (FNVA/AWU) and the family farm income per family working unit (FFI/FWU), which provide the best basis for comparisons across Member States because incomes per holding or per hectare are highly influenced by variations in farm size and type. However, the results should be treated with caution as the sample size is very small in certain Member States (Belgium, Spain, Portugal, and the United Kingdom). The small sample sizes also make it impossible to differentiate the EU results on either a regional or farm type basis.

2.2.4.3 Technology skills

IRENA 6 — Farmers' training levels

The Farm Structure Survey (FSS) data for the year 2000 at EU-15 level were used to give an indication of the level of agricultural training of farmers (defined as managers of agricultural holdings — category A). Data from the Common indicators for monitoring the implementation of rural development programmes have allowed identifying the share of training actions co-financed by the EAGGF-Guarantee fund (2001) aimed at preparing farmers for the application of production practices compatible with the protection of the environment and the maintenance and enhancement of the landscape.

2.2.4.4 Attitudes

IRENA 7 — Area under organic farming
This indicator provides trends in organic farming area, and in the share of organic farming area in total utilised agricultural area (UAA) at national level (1998–2002). The regional share of organic farming area in total utilised agricultural area (UAA) in year 2000 is provided at NUTS 2 or 3 level. At the national level, the underlying data come from a questionnaire managed by DG Agriculture and Rural Development and treated by Eurostat. Data is supplied by EU-15 Member States by using the statistical tables of the organic farming questionnaire (electronic version OFIS). Only the area certified under Regulation (EEC) No 2092/91 (sum of organic and in-conversion area) is taken into consideration.

Data is also supplied by EU-15 Member States to the Farm Structure Survey. This data has been used to calculate the regional share of organic farming area in the total UAA. However, in some cases, the regional data do not only cover organic farming areas certified by Regulation (EEC) No 2092/91, but also areas receiving agri-environment support for organic farming (e.g. in Sweden).

2.3 Indicator evaluation

Indicators are a key tool in agri-environmental reporting since they help to summarise and illustrate complex agri-environment relationships. Thus they facilitate the communication of research results. However, indicators are not a goal in themselves but should provide a meaningful contribution to environmental reporting. Successful agri-environmental indicators should contribute to the following objectives:

- simplified description of complex reality;
- better communication with non-specialists;
- analysis of environmental trends in longer time series;
- building a common basis for discussion; and,
- identifying priorities in political decision-making.

The following criteria, identified in COM (2001) 144, are used to evaluate the actual usefulness of individual agri-environmental indicators developed during the IRENA operation: policy relevance, responsiveness, analytical soundness, data availability and measurability, ease of interpretation, and cost effectiveness. A scoring scheme is devised for each criterion to evaluate the usefulness of each indicator (Table 2.1). High overall scores indicate more usefulness than lower scores.

The scores allocated to each (sub-) criterion give an overall score. The minimum possible score is 0; the maximum possible score is 20. Indicators are classified in three categories indicating different degrees of 'usefulness' or 'potential' of indicators on the basis of their overall score.

- **'Useful'** — indicators that score more than 14 points.
- **'Potentially useful'** — indicators that score 8 to 14 points.
- **'Low potential'** — indicators that score 7 points or less.

However, to be classified in the highest category ('useful') indicators have to show minimum scores for certain key criteria: 2 points for policy relevance, 4 points for analytical soundness, and 3 points for data availability and measurability. These thresholds were introduced to ensure that essential characteristics for indicators that support policy decisions are fulfilled. 'Potentially useful' indicators are those that can achieve the quality, which is required in policy decisions. Those characterised as 'Low potential' are not considered to merit further development.

When using the results of the indicator evaluation exercise it needs to be taken into account that even the most sophisticated scoring approach will not fully capture the complexities of agri-environmental analysis. While the framework used aims at weighting critical issues particularly highly, some indicators still achieve scores that do not seem to correspond to their usefulness or accuracy in an expert perspective. One example is the indicator on greenhouse gas emissions from agriculture (IRENA 19) where the crucial role of emission coefficients is not fully captured by the scores relating to data accuracy. Another example is the indicators that built on Corine land cover (IRENA 12 and 24) where potential biases in time trends inherent to the data sampling strategy do not have much weight in the final scoring. Overall, however, the evaluation scores give a good insight into the strengths and weaknesses of different indicators and indicator groups, in spite of individual shortcomings.

Table 2.1 Criteria used to evaluate individual agri-environmental indicators

Indicator criteria	Concept	Sub-criteria	Scoring
Policy relevance	Address the key agri-environmental issues	Is the indicator linked to Community policy targets, objectives or legislation?	0 = No
			1 = Yes, indirectly
			2 = Yes, directly
		Could the indicator provide information that is useful to policy action/decision?	0 = Not at all
			1 = Fairly useful
			2 = Very useful
Responsiveness	Changes sufficiently quickly in response to action	Is the indicator sensitive to changes in the phenomenon/process that it is meant to measure?	0 = Slow, delayed response
			1 = Fast, immediate response
Analytical soundness	Based on sound science	Is the indicator based on indirect (or modelled) or direct measurements of a state/trend?	0 = Indirect
			1 = Modelled
			2 = Direct
		Is the indicator based on low/medium/high quality statistics or data?	0 = Low quality statistics or data
			1 = Medium quality statistics or data
			2 = High quality statistics or data
		What are the causal links with other indicators within the DPSIR framework?	0 = Weak or no link
			1 = Qualitative link
			2 = Quantitative link
Data availability and measurability	Feasible in terms of current or planned data availability	Good geographical coverage?	0 = Only case studies
			1 = EU-15 and national
			2 = EU-15 national and regional
		Availability of time series	0 = No
			1 = Occasional data source (6)
			2 = Regular data source
Ease of interpretation	Communicate essential information in a way that is easy to understand for decision makers and the informed public	Are the key messages clear and easy to understand?	0 = Not at all
			1 = Fairly clear
			2 = Very clear
Cost effectiveness	Costs in proportion to the value of information derived	Based on existing statistics and data sets?	0 = No
			1 = Yes
		Are the statistics or data needed for compilation easily accessible?	0 = No
			1 = Yes, but requires lengthy processing
			2 = Yes

The scores are all reported in a number of indicator evaluation tables. The dark green colour in the last row of the tables marks all indicators that are classified as 'useful', light green represents 'potentially useful' indicators, and yellow denotes those considered to have 'Low potential'.

Agri-environmental indicators are not only evaluated in isolation but also in relation to other agri-environmental indicators. This appears essential in order to verify the usefulness of the entire set of indicators as well as of the underlying DPSIR model. To this end, the evaluation is carried out in the context of agri-environmental storylines that are developed by applying the DPSIR model, which explains why the same indicator can have different scores. Nearly all criteria evaluate the indicators according to current usefulness. In contrast, the evaluation of the extent to which an indicator provides information that is useful to policy action/decision (sub-criterion of the policy relevance criteria) was based on its potential usefulness, if conceptual limits and data constraints (or their insufficient spatial and temporal resolution) could be overcome.

---

(6) 'Occasional' data source is defined as short time series that are not derived from official statistics. 'Regular' data sources build on recognised statistics and normally stretch to a time series of at least ten years.

3 General trends in EU-15 agriculture

3.1 Summary of general trends in EU-15 agriculture

Cropping and livestock trends

- The utilised agricultural area (UAA) for EU-12 [7] decreased by 2.5 % (from 115.3 million ha to 112.7 million ha) between 1990 and 2000. Arable land decreased by 0.7 % (from 61.4 million ha to 61.0 million ha). Permanent grassland decreased by 4.8 % (from 43.5 million ha to 41.5 million ha). Permanent crops decreased by 3.8 % (from 10.3 million ha to 9.9 million ha).
- The number of livestock units of cattle decreased by 8.3 % between 1990 and 2000 (EU-12). The livestock units of sheep decreased by 3.4 % between 1990 and 2000 (EU-12). The livestock units of pigs, on the other hand, increased by 14.5 % between 1990 and 2000 (EU-12).

Trends in the intensity of farming and use of inputs

- In 1990, 44 % of the agricultural area of the EU-12 was managed by high-input farms, but this has decreased to 37 % in 2000. Low-input farms occupied the lowest share of the agricultural area in 1990 (26 %), but this share increased to 28 % in 2000.
- FADN data show considerable increases in milk yield (14 %) and cereal yield (16 %) between 1990 and 2000. This affects individual regions and farm types differently but is generally associated with better farm management, a targeted and sometimes high use of inputs and livestock feed, as well as advances in plant and livestock breeding and agricultural technology.
- In some regions the livestock stocking density increased by more than 10 %. Pig stocking density rose in Denmark, northern Germany, and north-eastern Spain. Sheep stocking density increased in southern Greece and central Spain. Cattle stocking density increased in southern France, southern Italy and western Spain.
- Total nitrogen (N) mineral fertiliser consumption in EU-15 decreased by 12 % from 1990–2001. Total phosphate (P_2O_5) mineral fertiliser consumption in EU-15 decreased by 35 % from 1990–2001 (running average).
- The total estimated amount of pesticides used in agriculture increased by 20 % between 1992 and 1999 (ECPA data) [8].
- The irrigable area in EU-12 increased from 12.3 million ha to 13.8 million ha between 1990 and 2000, representing an increase of 12 %. In France, Greece and Spain, the irrigable area increased from 5.8 million ha to 7.4 million ha between 1990 and 2000, representing an increase of 29 %.

Socio-economic trends in farming

- Between 1990 and 2000, the agricultural area in EU-12 managed by specialised farms has increased by 4 % (from 68.7 million ha to 71.2 million ha), whereas the area managed by non-specialised farms decreased by 18 % (from 33.7 to 27.7 million ha). The largest percentage change occurred on 'non-specialised livestock' farms, which declined by about 25 % in total area (from 15.8 to 11.9 million ha).
- Marginalisation — due to economic and demographic conditions — appears a risk in Ireland, the south of Portugal, Northern Ireland and large parts of Italy, leading to the possible abandonment of farming.
- The area under organic farming reached 3.7 % of the total UAA of EU-15 in 2002, up from only 1.8 % in 1998. Organic production accounted for 2 % of EU-15 total production of milk and beef in 2001, but less than 1 % of total production of cereals and potatoes.

[7] Belgium, Denmark, France, Germany, Greece, Ireland, Italy, Luxemburg, Portugal, Spain, the Netherlands, the United Kingdom.
[8] ECPA: European Crop Protection Association.

> **Land use changes**
>
> - During 1990 to 2000, the change in land use from agriculture to artificial surfaces ranged from 2.9 % in the Netherlands to 0.3 % in France. In general the highest percentage of agricultural land (in 1990) converted to artificial surfaces (by 2000) occurred in urban regions.
> - The NUTS regions with the largest percentage changes, and where agricultural land covered at least 150 000 ha in 1990, are Madrid (6 %), South Holland (5 %), and North Holland (5 %). Administrative regions in coastal areas also show significant changes in land use from agricultural land to artificial surfaces, such as: Alicante (3.6 %), Algarve (1.8 %) and Castellon (1.6 %). These changes are most likely linked to the growth of tourism.

3.2 Introduction

The importance of agriculture for the natural environment is emphasised by the fact that nearly half the land surface in EU-15 Member States is farmed. Farming is an activity, which goes beyond simple food production, affecting and using natural resources such as soil and water. Over the centuries, farming has contributed to the creation and maintenance of a large variety of semi-natural habitats and agricultural landscapes, and supports a diverse rural community that is an important European cultural asset.

European farming has changed dramatically during the last 50 years and will continue to change in the future. Technological developments, such as more efficient machinery, improved agrochemicals and seeds have allowed farmers to increase crop yields, and improved livestock breeds and feeding techniques have facilitated greater yields of milk and meat. EU and national financial support have underpinned this technological change in the drive to make farming more viable and competitive.

Table 3.1 shows the indicators employed in this chapter. Driving Force indicators are used to describe general trends in EU-15 agriculture during the 1990s: Energy use (IRENA 11), Land use change (IRENA 12), Cropping/livestock patterns (IRENA 13), Intensification/extensification (IRENA 15), Specialisation/diversification (IRENA 16) and Marginalisation (IRENA 17). This is supplemented by information on trends in the volumes of farm inputs used and the uptake of irrigation: Mineral fertiliser consumption (IRENA 8), pesticide consumption (IRENA 9) and water use (intensity) (IRENA 10). Complementary information includes indicators on organic farming (IRENA 5.1, 5.2 and 7), and the training levels of farmers (IRENA 6).

3.3 Trends in cropping and livestock patterns

Regional crop and livestock data from the Farm Structure Survey is used to determine trends in cropping and livestock patterns that provide insight into environmentally important trends in farming in the European Union. Major changes over time in

Table 3.1 IRENA indicators relevant for describing general trends in EU-15 agriculture

DPSIR	IRENA indicators	
Responses	No 5.1	Organic producer prices and market share
	No 5.2	Organic farm incomes
	No 6	Farmers' training levels
	No 7	Area under organic farming
Driving forces	No 8	Mineral fertiliser consumption
	No 9	Consumption of pesticides
	No 10	Water use (intensity)
	No 11	Energy use
	No 12	Land use change
	No 13	Cropping/livestock patterns
	No 15	Intensification/extensification
	No 16	Specialisation/diversification
	No 17	Marginalisation

General trends in EU-15 agriculture

Figure 3.1 Regional importance of the dominant agricultural land uses and the trend 1990–2000 ([9])([10])([11])

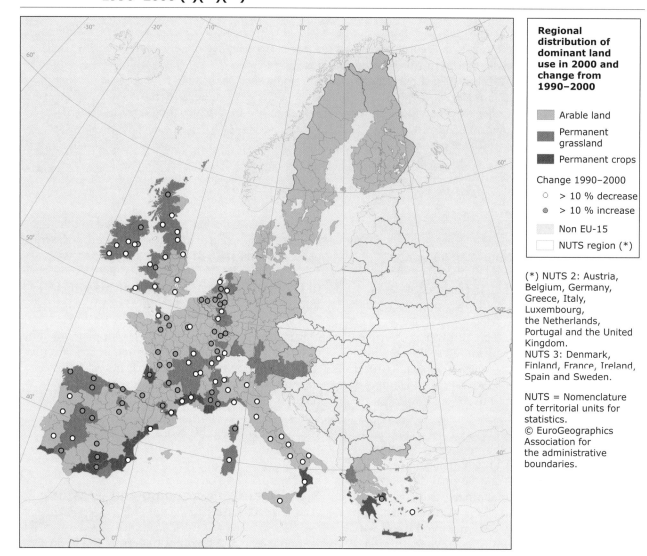

Source: Community survey on the structure of agricultural holdings (FSS), Eurostat.

the main land uses and livestock types may lead to positive or negative influences on the environment. The following sections highlight major regional trends as far as possible in the limited space available. Further information can be found in the underlying indicator fact sheets.

3.3.1 Cropping patterns

The utilised agricultural area (UAA) of the EU-12 decreased by 2.5 % (from 115.3 million ha to 112.4 million ha) between 1990 and 2000. There are large variations in the changes in UAA across the European Union. The largest changes in UAA are reported in Italy (– 3 121 910 ha, – 19 %) ([12]) and Spain (1 649 310 ha, + 7 %). Arable land decreased by 0.7 % (from 61.4 million ha to 61.0 million ha). The areas of permanent grassland and permanent crops decreased by 4.8 % (from 43.6 million ha to 41.5 million ha) and 3.8 % (from 10.3 million ha to 9.9 million ha), respectively.

([9]) Dominant land use is defined as the land use class with the largest area in the region concerned.
([10]) Trends are indicated for the areas where changes for the dominant class are higher than 10 %.
([11]) Information on trends in Finland, Sweden, and Austria and in the new Bundesländer in Germany is not available.
([12]) The reasons for the reported strong decline in UAA in Italy between 1990 and 2000 needs to be investigated on the basis of production and/or land use related data sets.

General trends in EU-15 agriculture

In most regions of the EU-15 arable land has the highest share in agricultural area of the three land use types displayed in Figure 3.1. In these regions the share of arable land is mostly stable, although it has decreased in parts of Italy. In parts of Ireland, the Netherlands, France and Spain, the proportion of arable land has increased. This is probably linked to an increased focus on forage crops rather than grassland in livestock production and the overall decline of cattle numbers. It should be noted that arable land can appear as dominant in Figure 3.1 at a share of 50 % or even less of agricultural area, depending on the size of the other land uses.

Regions dominated by permanent crops are found in Mediterranean countries (olives, fruit and wine production) and parts of France (mainly vineyards). The share of these crops as part of the total regional area has changed in many cases, but such trends are not consistent within individual Member States. The proportion of permanent grassland, in areas where this land use is dominant (mainly in the western EU-15), has decreased overall since 1990, with the exception of Spain. The largest decreases in permanent grasslands (more than 25 %) during the 1990s occurred in Denmark and central and western France (IRENA 13).

Figure 3.2 Regional distribution of dominant livestock types (expressed as livestock unit/ha UAA) and the change 1990–2000 ([13])([14])([15])

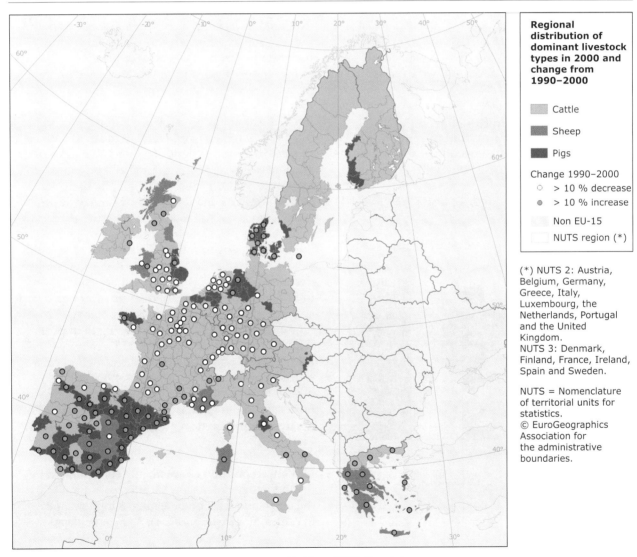

Source: Community survey on the structure of agricultural holdings (FSS), Eurostat.

---

[13] Trends are indicated for the regions where changes for the dominant class are larger than 10 % of 1990 levels.
[14] Information on trends in Finland, Sweden, and Austria and in the New Bundesländer in Germany is not available.
[15] Livestock unit (LU): Livestock numbers are converted into livestock units using coefficients available from the Eurostat Concepts and Definitions Database (Eurostat, 2004). These coefficients are provided in IRENA No 13 Cropping/livestock patterns.

3.3.2 Livestock patterns

Reported information on livestock numbers is standardised using livestock units to take into account the feeding regime for different livestock categories and age (Eurostat, 2004). At the European level livestock numbers expressed as livestock units were quite stable, decreasing by 1.9 % between 1990 and 2000 (EU-12). The number of livestock units of cattle decreased by 8.3 % between 1990 and 2000 (EU-12). Sheep livestock units decreased by 3.4 % between 1990 and 2000 (EU-12). The livestock units of pigs, on the other hand, increased by 14.5 % between 1990 and 2000 (EU-12).

Livestock patterns can also be expressed in livestock units per ha UAA, in order to indicate the stocking density and show the importance of a livestock type in a region. Figure 3.2 shows that cattle had the highest share of the total livestock population in many regions in the year 2000. In many cattle-dominated areas, cattle have declined by more than 10 %. Most areas where sheep farming and pig-rearing is dominant show an increase in sheep and pig livestock units, respectively.

3.4 Trends in the intensity of farming

3.4.1 Intensification and extensification

Intensification/extensification can be measured by changes in the number of livestock per area of land or the yield of selected crops considered in conjunction with the trends in the use of external inputs per cropped area. The Farm Structure Survey provides time series data on regional livestock numbers. Regional average yields for milk and major crops may be calculated based on FADN data. However, there are no regional data on the use of external inputs per cropped area. Instead trends in the regional average expenditure on farm inputs based on FADN data can be calculated ([17]). Changes in the share of agricultural land managed by three farm types are used as a proxy indicator (IRENA 15).

Intensification has been the predominant trend in most EU-15 regions for several decades. However, since 1990 there are signs of a trend towards a more efficient use of agricultural inputs, if measured by input costs recorded in FADN. The share of agricultural area managed by low- and medium-input farm types has increased slightly between 1990 and 2000. In 1990, low-input farms managed 26 % of the utilised agricultural area across EU-12, and this share increased to 28 % in 2000. High-input farms declined from 44 % of the utilised agricultural area to 37 % over the same period. Thus, although a large share of the agricultural area is still managed by high-input farms, they are decreasing in importance (Figure 3.3). Different types of farms compose the low, medium and high input categories and care needs to be taken in the interpretation of these overall figures which still hide variations between farm type and region. However, the approach allows the detection of some regional trends as shown in Figure 3.4.

Low-input farm types are mainly concentrated in the Iberian Peninsula, on Mediterranean islands, the north and west of the United Kingdom and in central France (Figure 3.4). As a generalisation,

Figure 3.3 Trends in the share of agricultural land managed by low-input, medium-input or high-input farm types ([16])

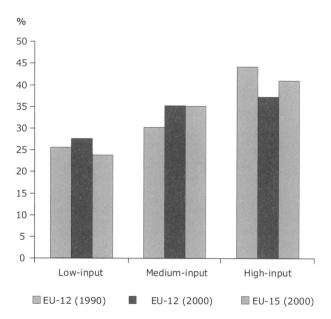

Source: FADN — DG Agriculture and Rural Development; adaptation LEI.

([16]) Farm types are defined as follows: Low-input farms spend less than 80 Euro per ha per year on fertilisers, crop protection and concentrated feedstuff. Medium-input farms spend between 80 and 250 Euro per ha per year and high-input farms more than 250 Euro per ha per year on these inputs.
([17]) These figures have to be taken as indication only as it was not feasible to ensure that differences in the costs of inputs between different Member States were fully harmonised in the data set used.

General trends in EU-15 agriculture

Figure 3.4 Regional importance of low-input, medium-input and high-input farming ([18]) and the trend 1990–2000 ([19])

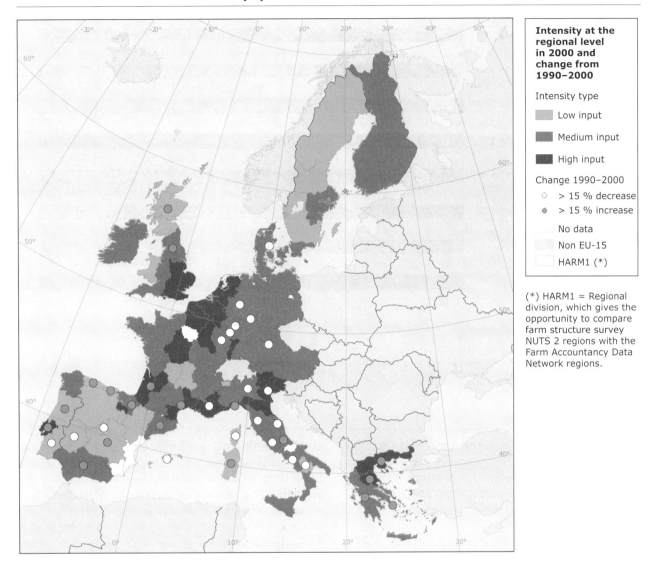

Source: FADN — DG Agriculture and Rural Development, adaptation LEI.

high input farm types are predominant in the Netherlands, Belgium, south-eastern England, northern France, northern Italy and northern Greece. Increasing trends in the expenditure on inputs can be identified in regions dominated by low input farm types, such as in the Mediterranean and Scotland. Significantly, several high input regions in Mediterranean countries and France also show an increase of input costs by more than 15 % between 1990 and 2000.

From an environmental perspective it is important to distinguish trends in input use between different farm types as many of them have specific environmental characteristics. Figure 3.5 shows trends in input costs for some relevant farm types, again based on FADN data. The overall decline in the share of high-input farms is particularly reflected in those farm types that have high external input use (cereals and mixed crops/livestock). For the mixed crops/livestock farms, the share of low-input farms has also increased considerably. This could reflect an increased efficiency of input use in the more intensive systems since 1990. The farm types that use lower inputs have changed less, with the share of high-input farms actually increasing slightly in the

---

[18] The low-input regions are the 20 regions with the lowest average expenditure on inputs; high-input regions are the 20 regions with the highest average expenditure on inputs, and medium-input regions constitute the remainder.
[19] Information on trends in Finland, Sweden, Austria, and in the New Bundesländer in Germany is not available.

General trends in EU-15 agriculture

Figure 3.5 Trends in intensity of farming for selected types of farms (derived typology) between 1990 and 2000 in EU-12

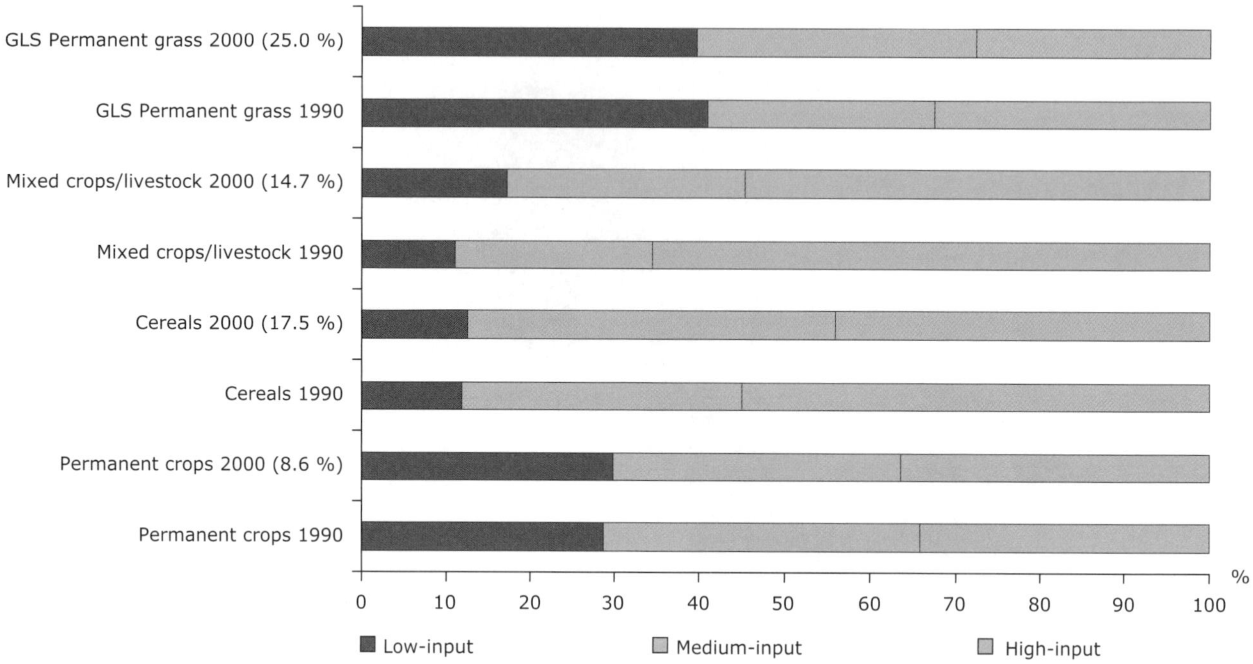

Farm type	Low-input	Medium-input	High-input
GLS Permanent grass 2000 (25.0 %)			
GLS Permanent grass 1990			
Mixed crops/livestock 2000 (14.7 %)			
Mixed crops/livestock 1990			
Cereals 2000 (17.5 %)			
Cereals 1990			
Permanent crops 2000 (8.6 %)			
Permanent crops 1990			

Note: The numbers in brackets indicate the area share of each farm type in total utilised agricultural area in 2000.

Source: FADN — DG Agriculture and Rural Development, adaptation LEI.

permanent crop systems. This could reflect changes in Mediterranean farm systems that are also evident in Figure 3.4. However, further work is required to analyse, for example, potential differential trends in input use intensity in different permanent crop systems (olive, fruit and wine production).

Additional information on intensification/ extensification trends can be derived from the development of milk and cereal yields (IRENA 15). FADN data show that average milk yields for the EU-12 increased by about 14 % between 1990 and 2000. This results from a higher use of protein-rich feed, advances in livestock breeding and more focused herd management. At the national level the strongest increases occurred in Portugal, Spain, Germany, Italy, Luxembourg and Greece. Milk production potential in the EU-15 increases from south to north due to natural conditions (length of grazing season, rainfall and temperature patterns). Figure 3.6 provides a picture of the regional distribution of these increases by FADN HARM regions ([20]). Increases above 15 % are mainly found in northern Italy, the northwest of Spain and Portugal, mountainous regions of France, Ireland, Belgium, the Netherlands, most of Germany and in Denmark. An analysis of milk yield increases among grazing livestock farm types shows that farms with a focus on permanent grassland increased the average yield by 17 % between 1990 and 2000 whereas farms where temporary pastures or forage crops (e.g. silage maize) are dominant achieved an increase of 22 % in milk yield.

The average increase in the yield of cereals for the EU-12 was 16 % between 1990 and 2000 (IRENA 15), with a slower increase towards the end of the decade. Figure 3.7 shows that yield increases occurred on all types of farms with the strongest increase on farms that specialise in cereal cropping. Improvements in farm management, a targeted and sometimes increased use of inputs, progress in plant breeding and technological advances, e.g. precision drilling, are key reasons for this yield increase. However, average cereal yields continue to vary strongly across the EU-15, with average yields of eight to ten tonnes per ha in favoured arable regions of the United Kingdom, Denmark, Germany or France and yields as low as two to three tonnes per ha in the dry interior of the Iberian peninsula.

([20]) The so-called HARM regions are constructed units that allow comparing Farm Structure Survey NUTS 2 regions with farm accountancy data network regions.

General trends in EU-15 agriculture

Figure 3.6 Regional distribution of milk yields in 2000 and change from 1990–2000

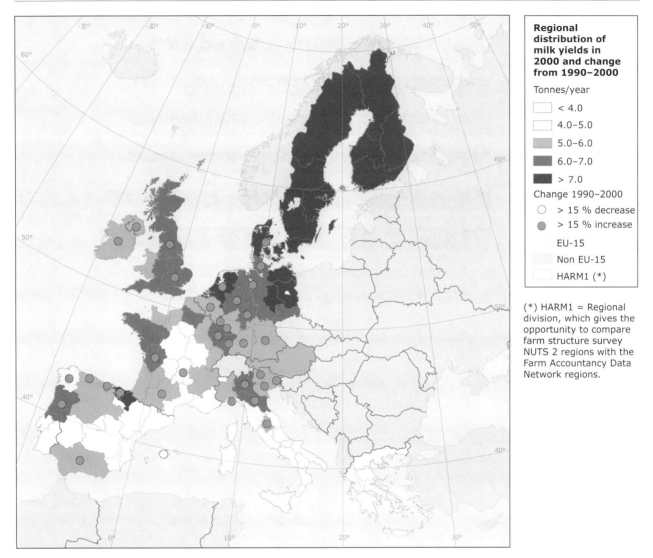

Note: Trends can only be shown for FADN regions with at least 15 sample farms.
Source: FADN — DG Agriculture and Rural Development, adaptation LEI.

The analysis of FADN data that underpins the IRENA 15 (intensification/extensification) points overall to a decrease of input costs coupled with a considerable increase of milk and cereal yields, both of which indicate a more efficient farm management. The use of farm types helps to understand certain patterns within the overall development but supplementary analysis has to be carried out to be able to link region and crop specific trends that ultimately determine potential environmental impacts.

3.4.1.1 Mineral fertiliser consumption

IRENA 8 provides the evolution of nitrogen and phosphate mineral fertiliser use over time based on Faostat data. Indicators of mineral fertiliser consumption show a declining trend:

- Total nitrogen (N) mineral fertiliser consumption in EU-15 decreased by 12 % from 1990 to 2001 (3-year averages). During this period, consumption decreased in most of the EU-15 Member States, except in Spain and Ireland. The biggest decreases (more than 30 %) occurred in Denmark and Greece.
- Total phosphate (P_2O_5) mineral fertiliser consumption in EU-15 decreased by 35 % from 1990 to 2001 (3-year averages). During the same period consumption decreased in all EU-15 Member States, except in Spain. The largest declines (more than 50 %) occurred in Germany, Denmark and Finland.

It is difficult to link these trends directly with environmental impact. The final effect on the

General trends in EU-15 agriculture

Figure 3.7 Changes in cereal yields for selected farm types ([21]) between 1990 (EU-12) and 2000 (EU-12)

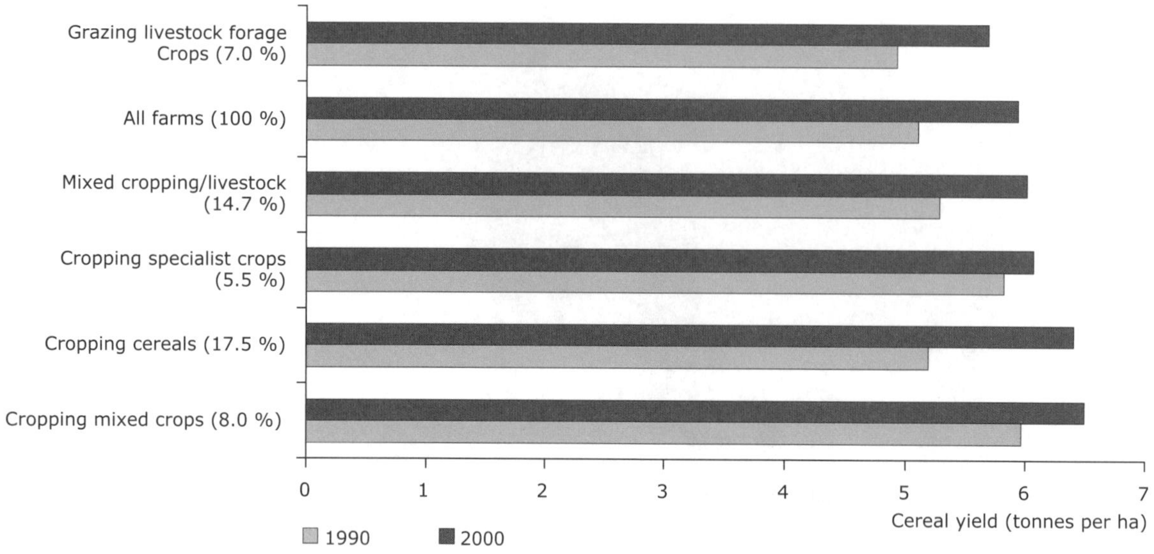

- Grazing livestock forage Crops (7.0 %)
- All farms (100 %)
- Mixed cropping/livestock (14.7 %)
- Cropping specialist crops (5.5 %)
- Cropping cereals (17.5 %)
- Cropping mixed crops (8.0 %)

Cereal yield (tonnes per ha)
■ 1990 ■ 2000

Note: The numbers in brackets show the share of agricultural area (%) managed by the farm type in 2000.
Source: FADN — DG Agriculture and Rural Development; adaptation LEI.

environment depends to large degree also on other factors, such as trends in the use of organic fertilisers, yields, cropped areas and farm management practices.

3.4.1.2 Consumption of pesticides

IRENA 9 provides information on pesticides sold (based on information from Member States) and pesticides used (based on information from ECPA ([22]). The total quantity of pesticides sold, expressed in active ingredient (a.i.), grew from 295 000 tonnes in 1992 to 327 000 tonnes in 1999, an increase of 11 %. Sales of fungicides and herbicides grew by 15 % and 11 % respectively, but sales of insecticides decreased by 16 %. However, sales figures also cover use outside agriculture.

The total estimated quantity of pesticides used grew from 194 000 tonnes a.i. in 1992 to 232 000 tonnes in 1999, representing an increase of 20 %, but these figures are significantly lower than sales volumes. Inorganic sulphur (a fungicide) represents a very large proportion of the quantities used.

The average estimated pesticide application rates (kg a.i./ha) are higher than the EU-15 average in Italy, Greece, Portugal and France. Average fungicide application rates (kg a.i./ha) are higher than the EU-15 average in Italy, Greece, Portugal, France and the Netherlands. This situation arises from the emphasis on sulphur fungicides used in vineyards in these Member States, except the Netherlands.

Currently, existing data does not allow an assessment of the potential increase in environmental risk associated with higher pesticide sales or use volumes. This is partly due to the lack of knowledge on the spatial, seasonal and crop application patterns of pesticides by farmers, and partly due to technical changes of the plant protection products themselves, in terms of active ingredients, application behaviour and decomposition patterns. However, a specific research project on harmonised pesticide risk indicators (HAIR) aims to provide a harmonised European approach for indicators of the overall risk of pesticides. Its implementation will require improved pesticide use data.

3.4.1.3 Water use (intensity)

The collection of actual water use (intensity) trends at the farm level is not feasible. Trends in the irrigable area (i.e. area equipped for irrigation) from the

---

[21] Farm types based on the Community Typology and certain land use criteria (Table A.3).
[22] European Crop Protection Association.

Figure 3.8 Regional map of the area of cultivated grain maize (2000) and the area of irrigated grain maize in France, Greece, Italy and Spain (2000)

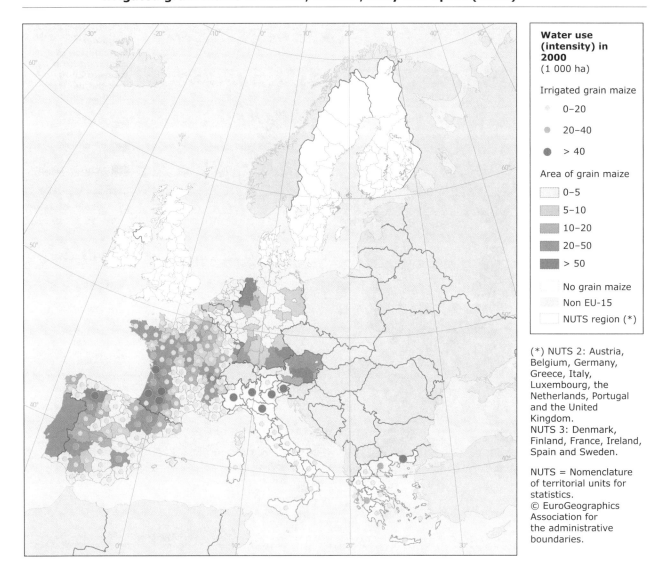

Source: Community survey on the structure of agricultural holdings (FSS), Eurostat.

Farm Structure Survey are therefore used as a proxy indicator (IRENA 10). Although actual irrigated area is generally lower than the irrigable area, information on irrigable area is collected in all EU-15 Member States. Expressing irrigable area as a percentage of utilised agricultural area gives an idea of the importance of irrigation in the agriculture sector.

The irrigable area in EU-12 increased from 12.3 million ha to 13.8 million ha between 1990 and 2000, representing an increase of 12 %. This is fully accounted for by France, Greece and Spain, where the irrigable area increased from 5.8 million ha to 7.4 million ha between 1990 and 2000, which represents an increase of 29 %. The irrigable area in EU-12, as a share of total utilised agricultural area, increased slightly from 10.3 % in 1990 to 11.7 % in 2000.

In southern Europe there is also information on selected crops that are irrigated at least once a year. The most important irrigated crop is grain maize. The area of irrigated grain maize increased by 23 % (0.3 million ha) between 1990 and 2000, mainly in France, Spain and Northern Italy (Figure 3.8).

3.4.1.4 Energy use

Direct energy use by the agriculture sector is linked mainly to the use of oil products and electricity for heating and fuel for farm machinery. Indirect energy use in agriculture is mainly for the production of fertilisers and pesticides, farm machinery and building. Total final energy consumption in agriculture as a percentage of total energy use in EU-15 Member States ranges between 0.5 % and 6.5 % (OECD, 2003).

General trends in EU-15 agriculture

Figure 3.9 Regional distribution of dominant farm types by specialisation and the trend 1990–2000 ([23])([24])

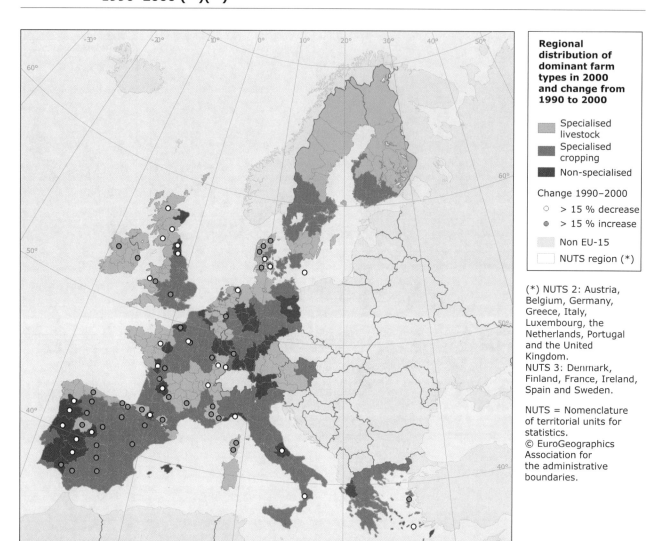

Source: Community survey on the structure of agricultural holdings (FSS), Eurostat.

Energy consumption per ha of UAA in EU-15 increased from 6 to 7 GJ/ha between 1990 and 2000 (excluding energy for fertiliser production). Crude oil and petroleum products are the main sources of energy consumption in agriculture in the EU-15. Motor fuels and lubricants account for more than half of total energy costs in most Member States, mainly for energy used in farm operations (e.g. ploughing, harvesting and drying). Natural gas is the main component of the large amount of energy used in agriculture in the Netherlands (65 GJ/ha), in particular for greenhouse production. In the EU-15, the share of inorganic fertilisers in total energy consumption is around 35 %. Estimates of energy used to produce inorganic fertilisers at Member State level enable a link to be made between energy use and agricultural area. Energy consumption related to inorganic fertilisers used in agriculture ranges from 9 GJ/ha (the Netherlands) to 2 GJ/ha (Portugal and Austria).

---

[23] 'Non-specialised' includes non-specialised livestock, non-specialised cropping and non-specialised cropping/livestock.
[24] Note that information on trends in the regions of Finland, Sweden, Austria and Germany is not available.

Figure 3.10 Share of gross farm income derived from agri-environment payments (2000)

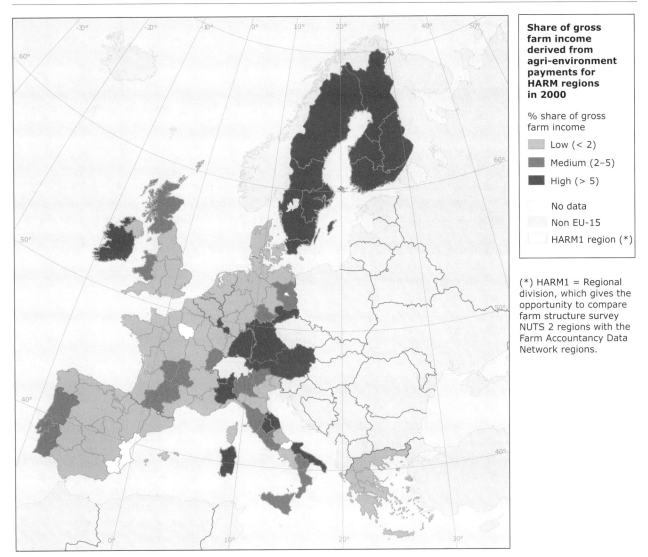

Source: FADN — DG Agriculture and Rural Development, adaptation to HARM regions by LEI.

3.5 Trends in specialisation and diversification

3.5.1 Specialisation

Specialisation occurs when a single type of production dominates farm income. For example, a mixed farmer could stop livestock farming to focus exclusively on arable production. The main forces behind this trend are the need for economic efficiency and changes in market conditions. Specialisation generally leads to a higher production efficiency (e.g. European Commission, 1999), but may also result in negative environmental effects. This is the case when it results in specialised, homogenous cropping or livestock patterns that eventually lead to a loss of diversity in farmland habitats, crop varieties and animal breeds. Serious environmental implications might result from the cumulative impact of such decisions over large areas. However, some specialised farming systems are linked to special agricultural landscapes. For example, extensive livestock farming in mountainous areas can be highly specialised, but it helps to maintain semi-natural grasslands and high nature value habitats.

The trend in the agricultural area managed by specialised types of farm can be used as an indicator of specialisation (IRENA 16). Between 1990 and 2000, the agricultural area in EU-12 managed by specialised farms has increased by 4 % (from 68.7 to 71.2 million ha), whereas the area managed by non-specialised farms decreased by 18 % (from 33.7 to 27.7 million ha). The most significant trend concerns 'non-specialised livestock' farms, which declined by 25 % in total area (from 15.8 to 11.9 million ha).

At regional level, the changes mainly concern the regions in which non-specialised farm types are predominant. Regions in Italy, Greece and Portugal have experienced large decreases in the share of agricultural area managed by non-specialised farm types. This means that more specialised farm types are gaining ground in these regions (Figure 3.9).

3.5.2 Farm diversification

Diversification of farms occurs when there is a widening of agricultural and non-agricultural activities on the farm, but can also refer to off-farm income generation (e.g. part-time labour) by farmers and/or family members. Farm diversification cannot be linked directly to environmental impacts, and diversification may not always affect farming practices. However, diversification usually stabilises farmers' incomes and may indirectly prevent farmland abandonment, which is usually considered environmentally undesirable. However, little data are available to monitor changes in farm diversification

The share of agri-environment payments to gross farm income can be used to assess to what extent farms are diversified towards delivering environmental services (IRENA 16). Increasingly, these payments have become a new source of income for farmers. This indicates maintenance of, or a possible change towards, environmentally friendly farming practices (e.g. extensive grazing, reductions in chemical inputs and maintenance of landscape features).

In 2000, the share of agri-environment payments in gross farm income was highest for specialised livestock farms, with an average contribution of 6.5 % (Figure 3.10), whereas they contribute on average only 3 % in specialised cropping farms. This may reflect the importance of grassland management as one important type of agri-environment measures supported.

The regional distribution of the share of agri-environment payments in gross farm income is highly variable. Agri-environmental payments are most significant in Sweden, Finland, Austria, Ireland and parts of Italy and Germany. In (parts of) these countries agri-environment payments contribute more than 5 % of gross farm income. Currently, therefore, agri-environment payments provide only a small share of total gross income of farmers although their importance for net income may be considerably larger. The extent to which farmers can participate in agri-environment schemes depends on the measures offered by Member States in their rural development programmes.

3.5.3 Farmers' training levels

Training allows farmers to become better equipped to deal with day-to-day farm management and to adapt more easily to new economic circumstances and new agri-environmental practices.

Farm structure survey data for the year 2000 at EU-15 level show that 83 % of farm managers held only practical experience, just 9 % received basic agricultural training and only 8 % received full agricultural training (IRENA 6). In 1990, these percentages were 87 %, 8 % and 5 %, respectively. However, Member States differ considerably in the level of agricultural training.

At EU-15 level, 14 % of the total number of training actions co-financed by the EAGGF-Guarantee fund ([25]) under rural development programmes (2001) were targeted at preparing farmers for environmentally friendly farming. Acquisition of skills needed to enable reorientation of production (47 %) and economic management (38 %) are still the objectives of the majority of training actions.

Information on training levels is not sufficiently targeted or reliable to draw strong conclusions about its implications for agri-environmental management on farms. A high level of agricultural training should facilitate, but does not guarantee, sound environmental management.

3.6 Trends in marginalisation and land use change

3.6.1 Marginalisation

Marginalisation is caused by low agricultural profitability, often linked to physical or climatic handicaps and wider socio-economic trends. Marginalisation can have far-reaching effects on the environment by favouring farm abandonment with an associated loss of biodiversity and heritage landscapes.

IRENA 17 uses FADN data to identify areas at risk of marginalisation by combining information on regions where farming is of low profitability and regions where many farmers are close to retiring age. Regions with low profitability are those where more than 40 %

---

[25] The European Agricultural Guidance and Guarantee Fund (EAGGF) finances training measures only outside the Objective 1 regions.

Figure 3.11 Change in land use from agriculture to artificial surfaces as a percentage of agricultural area (in 1990) mapped using a 3 km grid

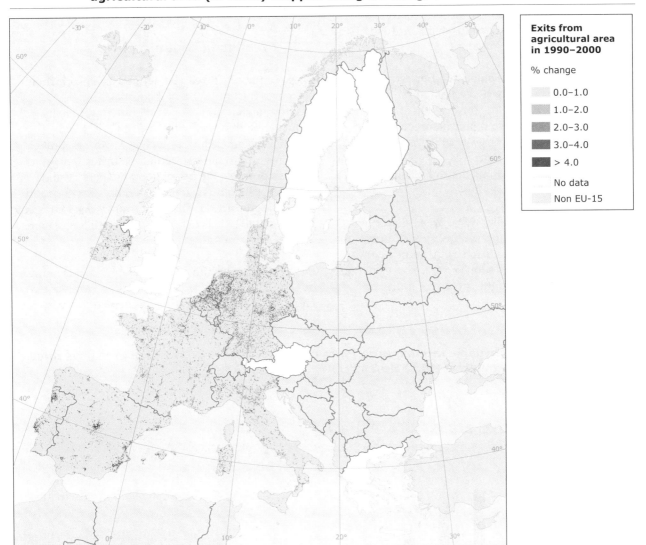

Source: Corine land cover.

of holdings have a farm net value added per annual work unit (FNVA/AWU) that is below 50 % of the average FNVA/AWU in that region. Regions with a high share of farmers close to retiring age are those where the proportion of holdings with farmers aged 55 years and over exceeds 40 %. Although the available data can hide significant intra-regional differences, it appears that marginalisation occurs in Ireland, the south of Portugal, Northern Ireland and large parts of Italy. Marginalisation seems to have increased during the 1990s in Northern Ireland and southern Portugal. FADN data and national information also point to the occurrence of marginalisation in parts of Spain and France (IRENA 17).

3.6.2 Land use change

The surface area devoted to agriculture is shrinking gradually in Europe, mainly due to afforestation and urbanisation. The satellite-based Corine land cover (CLC) 1990 and 2000 data represent the only European-wide database that can be used to identify changes from agriculture to artificial surfaces. At the end of 2004 CLC 2000 was available for Belgium, Denmark, France, Germany, Ireland, Italy, Luxembourg, the Netherlands, Portugal and Spain [26].

---

[26] Preliminary results are presented for France and Spain. The procedure for tracking land cover changes on the basis of CLC appears to cause under-reporting of actual changes. National data show a stronger urbanisation trend in France than detected by Corine land cover, while national and CLC data show the same trend in Germany.

General trends in EU-15 agriculture

IRENA 12 indicates the area of land use change from agriculture to artificial surfaces between 1990 and 2000, represented in absolute terms (hectares) and as a percentage of the agricultural area in 1990 (3 km grid, NUTS 2/3 regions, or country).

Land use changes represented at administrative level show that the highest percentage of agricultural land (in 1990) converted to artificial surfaces (by 2000) occurred in urban regions. The NUTS regions with the largest percentage changes, and where agricultural land covered at least 150 000 ha in 1990, are Madrid (6 %), South Holland (5 %), and North Holland (5 %).

Administrative regions in coastal areas also show significant changes in land use from agricultural land to artificial surfaces, such as: Alicante (3.6 %), Algarve (1.8 %) and Castellon (1.6 %). These changes are most likely linked to the growth of tourism.

detected, for example, between Paris and Brussels (Figure 3.11).

3.7 Trends in organic farming

Organic agriculture can be defined as a production system, which puts a strong emphasis on environmental protection and animal welfare by reducing or eliminating the use of GMOs and synthetic chemical inputs such as fertilisers, pesticides and growth promoters/regulators. Instead, organic farmers promote the use of good husbandry and agro-ecosystem management practices for crop and livestock production. The legal framework for organic farming in the EU is defined by Council Regulation 2092/91 and amendments.

Organic farming differs from conventional farming by the application of production rules, certification and a labelling scheme. This has helped to create a

Figure 3.12 Share of organic farming area (sum of organic and in-conversion area), certified under Regulation (EEC) No 2092/91, in total UAA

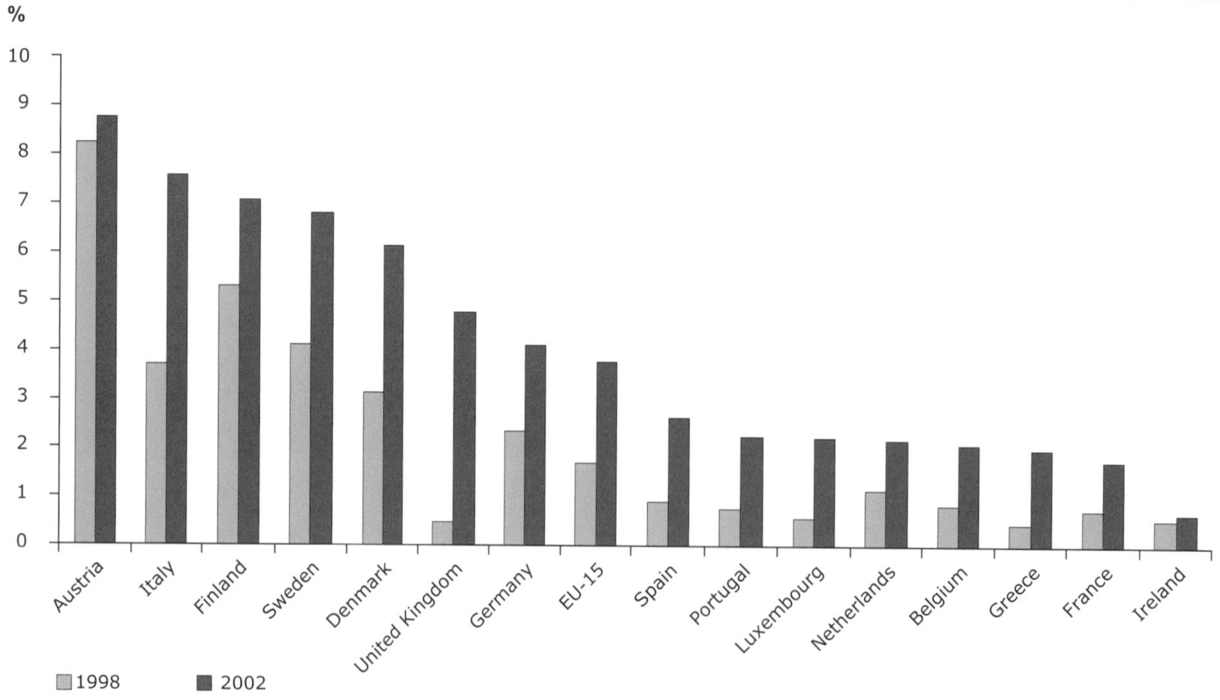

Note: The UAA total for Finland, Greece and the United Kingdom in 2002 is estimated on the basis of previous years.
Source: Organic farming questionnaire, DG Agriculture and Rural Development, data treated by Eurostat; ZPA1, Eurostat.

Land use changes represented at the 3km grid show more clearly the changes in coastal resorts. In addition, major surface transport axes that have been developed during the 1990s are also

distinct market, which is partially separated from non-organic produce. Agri-environment schemes have provided significant support for the expansion of organic farming (IRENA 7). The area under organic farming in 2002 covered 4.8 million ha in EU-15, an increase of 112 % compared to 1998 (IRENA 7). In 2002, the area under organic farming reached 3.7 % of total UAA in the EU-15, up from only 1.8 % in 1998. A quarter of the organic farming area in the EU-15 in 2002 was in Italy. The United Kingdom had the second largest area, followed by Germany, Spain and France. Member States with an increase in area under organic farming above or close to the EU-15 average were the United Kingdom, Luxembourg, Portugal, Belgium, Spain, France and Italy (Figure 3.12).

Apart from price premia the market share of organic products is a very good indicator of market development and consumer willingness to buy organic products (IRENA 5.1). The market share of organic food will also be a key factor for the future development of the sector.

In 2001, organic production accounted for 2 % of EU-15 total production of milk and beef, but less than 1 % of total production of cereals and potatoes. Organic food products accounted for 1–2 % of total EU-15 consumption, with organic beef and cereals having a higher share than milk and potatoes.

Ultimately, farm incomes will be the decisive factor for farmers to convert to or remain in organic farming (IRENA 5.2). EU-FADN ([27]) data for 2001 show that organic farms generate incomes comparable to those of conventional farms. In particular, returns to family and employed labour are similar, which is significant given the labour intensive character of organic farming.

3.8 Conclusions: evaluation of indicators

3.8.1 Summary: general evaluation

Nearly half (5 out of 13) ([28]) of the indicators used to show agricultural trends are classed as 'useful' — 'area under organic farming' (IRENA 7), 'land use change' (IRENA 12) and the farm trend indicators 'cropping/livestock patterns', 'intensification/intensification', 'specialisation/diversification' (IRENA 13, 15, and 16 respectively), while the rest is ranked as 'potentially useful'.

The following sections present in more detail the evaluation of individual indicators according to the criteria set out in Section 2.3. The overall scoring is summarised in Table 3.3.

3.8.2 Policy relevance

All indicators, except 'farmers' training levels' (IRENA 6), are considered to be directly or indirectly linked to particular Community targets, objectives or legislation. Those most directly linked to Community objectives or legislation are 'area under organic farming' (there are regulations on organic farming and an European action plan for organic food and farming), 'pesticides consumption' (there are several directives related to plant protection products, and the Commission is currently preparing a thematic strategy on the sustainable use of pesticides), and 'mineral fertiliser consumption' (linked to the nitrates directive).

As explained in Section 2.3, the evaluation of the extent to which an indicator provides information, which is useful to policy action/decision, was done according to its potential utility if conceptual limits and data constraints are overcome and not according to the current (actual) state of development.

The indicators which are considered to be the most useful to policy action/decision are those related to organic farming (IRENA 5.1, 5.2 and 7), 'water use (intensity)' (IRENA 10), and 'cropping/livestock patterns' (IRENA 13). Organic farming is interesting because it is a strongly developing production system within agriculture, which to a large degree is being driven by consumer demand. Organic farming has a Community legal framework supported by agri-environment support, and a European action plan for organic food and farming has been approved (2004). The Plan recognises the need to balance support for organic land management for environmental reasons with initiatives to support the development of the organic food market.

The trend in 'irrigable area' (IRENA 10) provides the only available information on demand for water from the agricultural sector. However, expansion in irrigable area does not necessarily result in a growing demand for water. The water framework directive (WFD) includes an objective on the good quantitative status of groundwater.

The indicator 'cropping/livestock patterns' provides information on important agri-environmental trends, such as the share of arable and grassland areas in total UAA. Trends in cropping/livestock patterns also provide relevant environmental information related to nutrient balances and soil cover.

([27]) Farm Accountancy Data Network.
([28]) Indicator 5 (organic farming) is considered as 2 indicators.

The indicator 'farmers' training levels' (IRENA 6) is considered as less policy relevant than the rest of the indicators. It may provide useful information for policy action: for instance, training support is a measure within rural development policy and is important tool for underpinning cross-compliance. However, the current definition (general level of training) has no obvious link with environmental farm management. A policy response indicator on training actions targeted on environmental management could be more appropriate.

3.8.3 Responsiveness

The indicators considered to reflect environmental, policy and economic changes in a relatively short time are organic farming prices and incomes (IRENA 5.1, 5.2), 'water use (intensity)' (IRENA 10), 'energy use' (IRENA 11), and 'cropping/livestock patterns' (IRENA 13). More time is needed for the other indicators to respond to external factors, e.g. for 'land use change' (IRENA 12).

3.8.4 Analytical soundness

All indicators are based on direct measurements — apart from 'marginalisation' (IRENA 17) and 'energy use' (IRENA 11), which combine different data sets and expert knowledge. The 'water use' indicator (IRENA 10) uses irrigable area as proxy information. The 'pesticide consumption' indicator is based on indirect measurements as the better available data set relates to pesticide sales, which is not equivalent to pesticide consumption in agriculture.

High quality statistics or data are used to underpin most indicators, apart from 'organic producer prices/income' (5.1, 5.2), 'mineral fertiliser consumption' (IRENA 8), 'pesticide consumption' (IRENA 9), which are not underpinned by 'official statistics'.

Concerning the sub-criterion on links with other indicators, only 'cropping/livestock patterns' (IRENA 13) is considered to have a strong causal quantitative link with the description of general farming trends. The remainder have qualitative links with the other indicators but are difficult to relate in quantitative terms. 'Farmers' training levels' (IRENA 6) and 'energy use' (IRENA 11) have weak links with other indicators.

3.8.5 Data availability and measurability

As regards geographic coverage, all indicators are based on regional data, apart from 'organic producer prices/income' (5.1, 5.2), 'mineral fertiliser consumption' (IRENA 8), and 'pesticide consumption' (IRENA 9), which are based on national data.

Regular data sources providing long term data series exists for all the indicators apart from 'organic producer prices/income' (5.1, 5.2), which is based on occasional data sources.

3.8.6 Ease of interpretation

All indicators used to describe the trends in agriculture provide messages that are very clear to understand apart from 'farmers' training levels' (IRENA 6) since the information it provides has little relevance to environmental issues, 'pesticide consumption' (IRENA 9) and 'marginalisation' (IRENA 17). In the case of marginalisation, the indicator combines social and economic data to derive the share of farms, which are at risk of marginalisation, which has only a qualitative link to possible farm abandonment. The indicator on 'pesticide consumption' is not easy to understand because there are large discrepancies between 'sales' and 'use' data.

3.8.7 Cost effectiveness

Cost effectiveness is evaluated according to the current existence of statistics or data sets to underpin the indicators, as well as their accessibility and processing requirements.

All indicators are based on existing statistics and data sets, apart from 'organic farming prices' (IRENA 5.1), which is reliant on a research project (OMIARD [29]).

All underpinning data are considered to be easily accessible, with the exception of those not based on regular statistics (IRENA 5.1 and IRENA 5.2). There are some indicators for which data can be easily accessed but which require considerable processing to obtain results. This is the case for 'land use' (IRENA 12) — which needs substantial manipulation of Corine land cover data. The indicators 'intensification/extensification' (IRENA 15), 'specialisation' (IRENA 16) and 'marginalisation' (IRENA 17) are based on a combination of different data extracted from FADN and FSS.

---

[29] 'Organic marketing initiatives and rural development'.

Table 3.3 Evaluation of IRENA indicators used to analyse general trends in agriculture

Indicator criteria	Sub-criteria	Scoring	General trends in agriculture												
			Organic farming prices	Organic farming incomes	Farmers training level	Area under organic farming	Mineral fertiliser consumption	Consumption of pesticides	Water use (intensity)	Energy use	Land use change	Cropping/livestock patterns	Intensification/extensification	Specialisation/diversification	Marginalisation
		IRENA indicator no	5.1	5.2	6	7	8	9	10	11	12	13	15	16	17
Policy relevance	Is the indicator directly linked to Community policy targets, objectives or legislation?	0 = No 1 = Yes, indirectly 2 = Yes, directly	1	1	0	2	1	2	1	1	1	1	1	1	1
	Could the indicator provide information that is useful to policy action/decision?	0 = Not at all 1 = Fairly useful 2 = Very useful	2	2	1	2	1	1	2	1	2	2	1	1	1
Responsiveness	Is the indicator responsive to environmental, economic or political changes?	0 = Slow, delayed response 1 = Fast, immediate response	1	1	0	0	0	0	1	0	0	1	0	0	0
Analytical soundness	Is the indicator based on indirect (or modelled) or direct measurements of a state/trend?	0 = Indirect 1 = Modelled 2 = Direct	2	2	2	2	2	0	0	1	2	2	2	2	1
	Is the indicator based on low/medium/high quality statistics or data?	0 = Low quality statistics/data 1 = Medium quality statistics/data 2 = High quality statistics/data	1	1	2	2	1	1	2	2	2	2	2	2	2
	What are the causal links with other indicators within the DPSIR framework?	0 = Weak or no link 1 = Qualitative link 2 = Quantitative link	1	1	0	1	1	1	1	0	1	2	1	1	1
Data availability and measurability	Good geographical coverage?	0 = Only case studies 1 = EU-15 and national 2 = EU-15, national and regional	1	1	2	2	1	1	2	2	2	2	2	2	2
	Availability of time series	0 = No 1 = Occasional data source 2 = Regular data source	1	1	2	2	2	2	2	2	2	2	2	2	2

General trends in EU-15 agriculture

Indicator criteria	Sub-criteria	Scoring	General trends in agriculture												
			Organic farming prices	Organic farming incomes	Farmers training level	Area under organic farming	Mineral fertiliser consumption	Consumption of pesticides	Water use (intensity)	Energy use	Land use change	Cropping/livestock patterns	Intensification/extensification	Specialisation/diversification	Marginalisation
		IRENA indicator no	**5.1**	**5.2**	**6**	**7**	**8**	**9**	**10**	**11**	**12**	**13**	**15**	**16**	**17**
Ease of interpretation	Are the key messages clear and easy to understand?	0 = Not at all 1 = Fairly clear 2 = Very clear	2	2	1	2	2	1	2	2	2	2	2	2	1
Cost effectiveness	Based on existing statistics and data sets?	0 = No 1 = Yes	0	0	1	1	1	1	1	1	1	1	1	1	1
	Are the statistics or data needed for compilation easily accessible?	0 = No 1 = Yes, but requires lengthy processing 2 = Yes	1	1	2	2	2	2	2	1	1	2	1	1	1
Total score			**13**	**13**	**13**	**18**	**14**	**12**	**16**	**13**	**16**	**19**	**15**	**15**	**13**
Classification: 0 to 7 (*) = 'Low potential' 8 to 14 (**) = 'Potentially useful' 15 to 20 (***) = 'Useful'			**	**	**	***	**	**	***	**	***	***	***	***	**
Final classification of Indicators according to the following criteria: Policy relevance at least 2 points, Analytical soundness at least 4 points, Data availability at least 3 points			**	**	**	***	**	**	**	**	***	***	***	***	**

4 Agricultural water use

4.1 Summary of main points

- During the 1990s, the reported water allocation rates for irrigation decreased across the EU-15 Member States. This indicates a likely reduction in water application rates per hectare of land irrigated implying an increase in water use efficiency.
- The demand for irrigation water shows a strong regional distribution. From a total of 332 regions, the 41 regions with the highest use of water for agricultural purposes (more than 500 million m³/year) are all located in southern Europe ([30]).
- The limited data available indicate that the share of agriculture in water use remained stable in the period 1991–1997 in both northern and southern EU-15 countries, at about 7 % and 50 %, respectively.
- Water use issues are addressed in the codes of good farming practice of all southern Member States, but little information is available on the use of agri-environment schemes for this purpose.
- Data for many indicators in this storyline are patchy or not available. Thus, the links between the different indicators could only be explored tentatively.

4.2 Introduction

Agriculture is an important sector in terms of total water usage in Europe. New production methods reliant on irrigation play an important role in the development of the agricultural sector in many Member States, but the increase of agricultural irrigation can put pressure on water resources. The problem is exacerbated during prolonged dry periods such as the drought in the Iberian Peninsula between 1990 and 1995. Water availability problems occur when the different competing demands for water exceed average annual supply (e.g. EEA, 2003a).

The factors that may drive changes in water use for irrigation are:

- Relative yields for irrigated versus non-irrigated crops;
- Relative subsidy levels for irrigated versus non-irrigated crops;
- Large water supply projects (e.g. dams);
- Innovations in irrigation technology;
- Irrigation investment costs;
- Water prices.

Farmers may select crops that are more sensitive to water stress than others, for example potatoes and sugar beet in northern Europe, and vegetables or grain maize in southern Europe. In addition, crops that are very water-intensive may be selected, such as cotton or rice. Soil types influence the decision to invest in irrigation or not, especially in northern Europe (light sandy soils retain less water than other soils). The provision of irrigation infrastructure via large water supply projects gives farmers the option to adopt irrigation agriculture and produce different crops that are not possible under local rainfall conditions.

Innovations in irrigation technology that reduce the operational costs of irrigation may encourage farmers to embark on irrigated agriculture. However, installing irrigation equipment represents a major capital outlay for farmers, and farmers will expect to recover irrigation equipment investments by achieving higher returns. Farmers can achieve this by increased yields, producing higher value crops, or receiving increased subsidies, e.g. for irrigated maize. The price of water is an important running cost (especially if full costs are paid by the user) that the farmer will have to take into account when considering investment in irrigation.

An increase in the irrigated area in a Member State or region could lead to an increase in water

---

([30]) Northern EU-15 comprises Austria, Belgium, Denmark, Finland, Germany, Ireland, Luxembourg, the Netherlands, Sweden and the United Kingdom. Southern EU-15 comprises France, Greece, Italy, Portugal and Spain.

Agricultural water use

use for agriculture, unless improvements in efficiency can keep total water consumption at previous levels. Increasing water abstraction rates may give rise to environmental problems (Baldock *et al.*, 2000), such as:

- lowered water tables and lower river flows;
- salinisation or contamination of groundwater;
- conversion of extensive farmland (e.g. pseudo-steppes) to irrigated fields;
- secondary effects, which are much more difficult to measure, such as the disappearance of wetlands;
- damage to terrestrial and aquatic habitats upstream due to installation of dams and reservoirs.

4.3 IRENA indicators related to water resources

This chapter examines the main factors in the relationship between agriculture and water resources.

The Driving force — Pressure — State/impact — Response framework provides a means of showing linkages and associations between indicators (Figure 4.1 and Table 4.1) and helps to structure the environmental assessment of agriculture's impacts on water resources. Analysing this relationship is a difficult task, however, due mainly to the shortage of data on various key variables concerning water resource management.

Figure 4.1 Agricultural water use according to the DPSIR framework

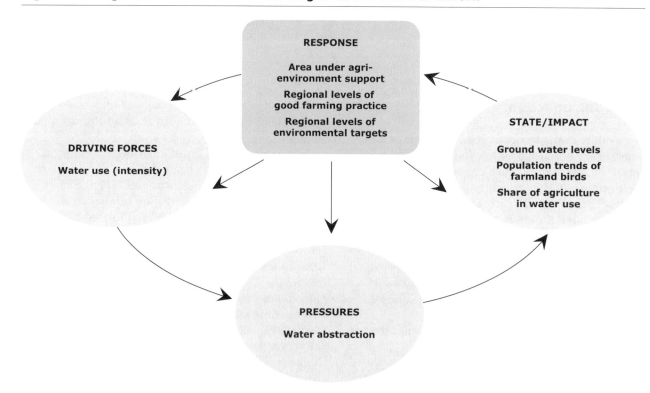

Table 4.1 IRENA indicators relevant for assessing agricultural water use

DPSIR	IRENA indicators	
Driving forces	No 10	Water use (intensity)
Pressures	No 22	Water abstraction
State	No 31	Ground water levels
Impact	No 28	Population trends of farmland birds
	No 34.3	Share of agriculture in water use
Responses	No 1	Area under agri-environment support
	No 2	Regional levels of good farming practice

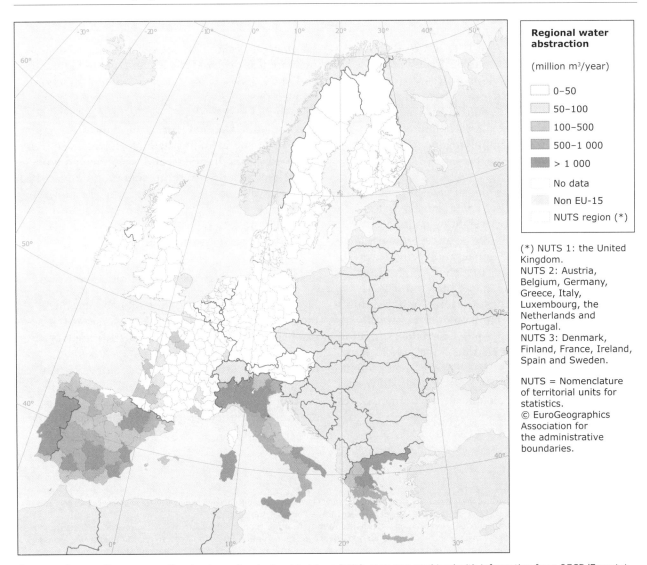

4.4 Agricultural driving forces

The main agricultural driving force behind the use of water is the consumption of water for irrigation. The key results of the indicator 'water use (intensity)' (IRENA 10) have been described in Chapter 3.

The irrigable area in EU-12 increased from 12.3 million ha to 13.8 million ha between 1990 and 2000, an increase of 12 %. This is fully accounted for by France, Greece and Spain, where the irrigable area increased from 5.8 million ha to 7.4 million ha during the same timeframe, representing an increase of 29 %.

4.5 Agricultural pressures on water resources

An increase in irrigable area may potentially have an impact on the demand for water because more farmers are likely to use irrigation methods. However, the adoption of improved irrigation technology, for example, from sprinkler to drip systems, will improve the water use efficiency of irrigation systems, reducing gross water requirements.

Several measures can support the increase of water use efficiency or limit excessive water use, including water pricing, use restrictions or running costs. However, in general the conversion from rain-fed to

Figure 4.2 Regional water abstraction rates for agriculture (million m³/year) during 2000 ([31])

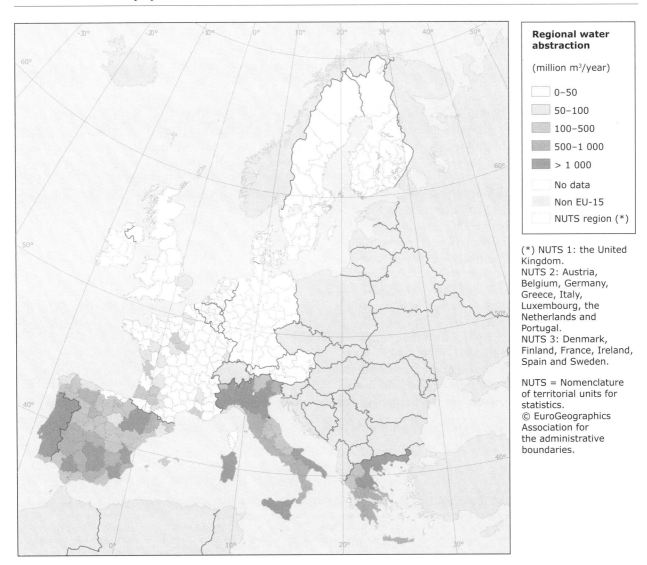

(*) NUTS 1: the United Kingdom.
NUTS 2: Austria, Belgium, Germany, Greece, Italy, Luxembourg, the Netherlands and Portugal.
NUTS 3: Denmark, Finland, France, Ireland, Spain and Sweden.

NUTS = Nomenclature of territorial units for statistics.
© EuroGeographics Association for the administrative boundaries.

Source: Community survey on the structure of agricultural holdings (FSS), Eurostat combined with information from OECD/Eurostat questionnaire.

[31] United Kingdom estimations are based on 1997 data for irrigable area and reported water abstraction rates. Ireland, Luxembourg and Germany do not provide data on irrigable area for NUTS regions.

Agricultural water use

irrigated agriculture will have a profound effect on water resources. Weather patterns will determine the annual demand for irrigation, in situations where irrigation is used to supplement rain-fed agriculture.

IRENA 22 provides national water allocation rates for irrigation based on annual abstraction rates (source OECD/Eurostat questionnaire) and irrigable area (Farm Structure Survey). The reported annual water allocation rates for irrigation are grouped into two regions: northern and southern EU-15 Member States ([32]). In northern EU-15 Member States, the reported mean annual water allocation rates decreased from 757 to 349 m^3/ha/year between 1990 and 2000. During this period the reported water abstraction decreased from 1 622 million m^3/year to 716 million m^3/year, and the irrigable area decreased from 2.1 million to 2.0 million ha.

In southern EU-15 Member States, the mean annual water allocation rates declined from 6 578 to 5 500 m^3/ha/year between 1990 and 2000. During this period the reported water abstraction rates decreased from 69 103 million m^3/year to 66 424 million m^3/year, whereas the irrigable area increased from 10.5 million ha to 12 million ha. This indicates a likely reduction in water application rates per hectare of land irrigated implying an increase in water use efficiency.

The IRENA 22 sub-indicator estimates regional water abstraction rates (see comment above) for agriculture, calculated by weighting national reported water abstraction rates by regional irrigable area (Figure 4.2). This regionalisation provides a good indication of regions that have a high water demand among the 332 regions analysed. The 41 regions with the highest use of water for agricultural purposes (more than 500 million m^3/year) are all located in southern Europe ([33]). Given the estimation method it is not possible to draw direct conclusions on water use intensity per ha of land in the different region from these figures, but the analysis shows the spatial distribution of potential abstraction pressures across the EU-15.

In general, the statistical information on irrigable area (from FSS) is more complete than the reported annual water abstraction for agriculture (from the joint OECD/Eurostat questionnaire). The data quality of the derived indicator is similar, therefore, to that of the water abstraction data.

4.6 State of/impacts on water resources

4.6.1 Groundwater levels

One impact of water demand that exceeds water supply is a progressive depletion of surface water and groundwater resources. However, data on groundwater levels (IRENA 31) in the EU-15 Member States are scarce. IRENA 31 is therefore based on a case study to illustrate the impact of changing water demands on groundwater levels.

La Mancha Occidental, Upper Guadiana basin in Spain was declared to be overexploited at the end of the 1980s. Unsustainable water abstraction had led to a severe negative impact on the nature reserve and RAMSAR ([34]) and Natura 2000 site of 'Las Tablas de Daimiel', threatening the destruction of this wetland area. Restrictions on water use were imposed during the 1990s and implemented with the help of an agri-environmental programme which reduced water abstraction rates from 600 million m^3 per year (in 1986) to 300 million m^3 per year (in 1996) (Figure 4.2). A steady recovery of regional groundwater levels resulted (EEA, 2003a).

4.6.2 Population trends of farmland birds

Negative impacts of irrigation on habitats and biodiversity can arise from the conversion of extensive farmland to irrigated agriculture. For example, cereal steppe bird habitats linked to dryland agriculture are largely eliminated when irrigation is introduced (Heath and Evans, 2000). Further consequences arise from a higher use of agricultural inputs to increase agricultural returns

---

[32] Northern EU-15 comprises Austria, Belgium, Denmark, Finland, Germany, Ireland, Luxembourg, the Netherlands, Sweden and the United Kingdom. Southern EU-15 comprises France, Greece, Italy, Portugal and Spain.

[33] In southern Europe there are twenty-one regions estimated to require more than 1000 million m^3/year. In Greece these regions are: Anatoliki Makedonia (Thraki), Kentriki Makedonia and Thessalia (representing 58 % of Greek agricultural water abstraction). In Italy these regions are: Piemonte, Lombardia, Veneto, Emilia-Romagna, Puglia, Sicilia and Sardegna (representing 75 % of Italian agricultural water abstraction). In Spain these regions are: Sevilla, Jaén, Ciudad Real, Valencia, Murcia, Huesca and Zaragoza (representing 40 % of Spanish agricultural water abstraction). The Portuguese regions are: Norte, Centro, Alentejo, and Lisboa e vale do Tejo.

[34] Convention on Wetlands, signed in Ramsar, Iran, in 1971, is an intergovernmental treaty, which provides the framework for national action and international cooperation for the conservation and wise use of wetlands and their resources.

Figure 4.3 Annual abstractions from the aquifer (left) and water-level recovery (right) — La Mancha Occidental, Upper Guadiana basin

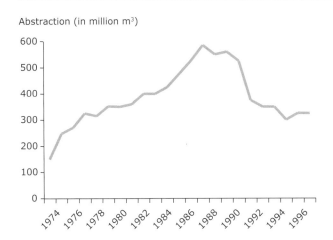

Source: Spanish Ministry of Environment, 2000.

and/or a change in the local water table regime. The increased use of agricultural inputs in irrigation schemes may affect a range of species, including farmland birds and aquatic organisms. Alterations to the local water table regime as a result of large water quantities abstracted from rivers or reservoirs may affect riparian habitats. However, information on declining farmland bird population trends (IRENA 28) currently cannot be linked specifically to the presence or introduction of irrigation schemes.

4.6.3 Share of agriculture in water use

The share of agriculture in water use is based on data from the OECD/Eurostat questionnaire (years 1990 and 1998) (IRENA 34.3). Data were calculated as three-year averages for two regions: northern and southern EU-15 Member States ([35]). The limited data available indicate that the share of agriculture in water use remained stable in the period 1991–1997 in both northern and southern Europe, at about 7 % and 50 %, respectively

4.7 Responses

4.7.1 Regional levels of good farming practice

The introduction of codes of good farming practice (GFP) is a policy measure to encourage the promotion of better management practices (IRENA 2) that will enable, amongst other objectives, the improvement of irrigation practices. Member States have to define codes of good farming practice at national or regional level in their rural development programmes (RDPs). Adherence to GFP is a basic condition for receipt of agri-environment and Less Favoured Area support. The objective is to encourage a better management of water resources at farm level and to support the enforcement of legislation on water. By providing environmental baselines the codes of GFP also help to ensure that agri-environment schemes deliver more environmental benefits throughout the EU.

Good farming practices in relation to irrigation methods and equipment are addressed in the codes of Spain, Greece, Portugal and France ([36]) where the scale of irrigation is significantly greater than in northern countries (Table 4.2). These include local regulations setting up conditions for water abstraction, making surface water abstraction subject to authorisation or declaration, and the monitoring of groundwater pumping by obligatory installation of counter systems at the withdrawal points (e.g. in France).

4.7.2 Area under agri-environment support

Agri-environment measures aim to support better environmental management by farmers across a range of environmental issues. However, the

---

([35]) Northern EU-15 comprises Austria, Belgium, Denmark, Finland, Germany, Ireland, Luxembourg, the Netherlands, Sweden and the United Kingdom. Southern EU-15 comprises France, Greece, Italy, Portugal and Spain.
([36]) Germany also has provisions relating to agricultural water use in its national water law.

Agricultural water use

Table 4.2   Degree of coverage of water-use and irrigation practices by national codes of GFP

Farming practices	BE-Fl	BE-Wa	DK	DE	GR	ES	FR	IE	IT-ER	LU	NL	AT	PT	FI	SE	UK
Water use: irrigation	—	—	—	—	■	■	■	—	—	—	—	—	■	—	—	—

■ Priority issue   — Issue not covered   ☐ Issue addressed

Source: Compiled from codes of GFP described in national rural development programmes 2000–2006.

information available at EU level about extent and purpose of agri-environment schemes does not allow identifying the number of schemes targeted at improving the efficiency of water use. However, the case study on the area of 'Las Tablas de Daimel' (south-central Spain) is an example of an agri-environment scheme that was implemented to reduce the impact of irrigation on the environment and biodiversity. A scheme was introduced in 1993 under which farmers in the area received payments in return for converting to crops that are less water demanding. As a result there has been a notable shift in cropping patterns (e.g. from sugar beet and maize to wheat). The scheme was successful in reducing abstraction rates, but at a high cost (Sumpsi et al., 2000).

4.8   Conclusions: evaluation of indicators

4.8.1 Summary: general evaluation

Six out of the seven indicators used to describe the driving forces and pressures related to agricultural water use, the state of and impact on water resources and responses have been evaluated as 'potentially useful'. The indicator 'groundwater levels' (IRENA 31) is considered to have low potential, mainly because the necessary data are not readily available.

In spite of a high score for several criteria, such as policy relevance and measurability, the water use indicator (IRENA 10) is classed in the category 'potentially useful' because trends in the irrigable area are only a proxy indicator for water use intensity. All the other indicators, apart from ground water levels, provide qualitative input into the environmental storyline, and information is only available at the national level.

The two response indicators 'area under agri-environment support' (IRENA 1) and 'good farming practices' (IRENA 2) are also considered as being potentially useful.

The following sections present in more detail the evaluation of individual indicators according to the criteria set out in Section 2.3. Table 4.3 summarises the scoring for all indicators in this storyline.

4.8.2 Policy relevance

All the indicators are directly or indirectly linked to particular Community targets, objectives or legislation, apart from 'share of agriculture in water use' (IRENA 34.3). The indicator on groundwater levels (IRENA 31) is directly linked to the water framework directive, as the maintenance of a good quantitative status of groundwater is one of its objectives. The indicator on population trends of farmland birds (IRENA 28) is considered relevant to policy in the context of the 2010 biodiversity target.

The water use (intensity) (IRENA 10) indicator is considered to be useful to policy makers' action/decision as it directly shows regional trends in the expansion of irrigable areas. It indicates where pressures on water resources are likely to occur — especially if demands from other sectors are known. Despite having a low potential due to the absence of data, the indicator on 'groundwater levels' (IRENA 31) is considered to be useful for policy action/decision. The response indicators (IRENA 1 and 2) are also very useful as they show which measures are being taken to improve water management (practices).

4.8.3 Responsiveness

The indicators that are sensitive to economic and policy changes are 'water use (intensity)', 'water abstraction rates' and 'groundwater levels'. If market conditions change, farmers may quickly adopt irrigation (but only if there is water provision), for example in the case of growing vegetables or fruits. The conversion to irrigation will have an immediate effect on water abstraction rates and groundwater levels. None of the other indicators are regarded as being particularly sensitive to environmental, economic or political changes.

4.8.4 Analytical soundness

All indicators are based on direct measurements, apart from the water use indicator (IRENA 10), as irrigable area is a proxy indicator of quantity of water use. Only 'water use (intensity)' (IRENA 10) is based on 'high quality' statistics or data.

The indicators 'water abstraction rates' (IRENA 22) and 'share of agriculture in water use' (IRENA 34.3) are underpinned by 'medium quality' data, as they are based on the joint OECD/Eurostat questionnaire. There are many gaps in the data and annual updates of information are not always provided. The 'population trends of farmland birds EU-15' (IRENA 28) is considered as medium quality data as bird counts cannot be as accurate as statistical surveys.

At present the only indicators with qualitative links with other indicators within the DPSIR framework are the indicators 'water use (intensity)', 'population trends of farmland birds' and the two response indicators. There are no indicators with strong quantitative links. 'Water abstraction rates' (IRENA 22) and 'share of agriculture in water use' (IRENA 34.3) would have strong quantitative links in the DPSIR framework if high quality data becomes available. This is because it would be possible to monitor trends in the use of water resources, which would enable better targeting of measures to improve the long-term availability of water resources.

4.8.5 Data availability and measurability

Only the water use (intensity) indicator (IRENA 10) is based on regional and regular (long term series) data. All the pressure, State/impact and response indicators are based on national information, except the groundwater levels indicator (IRENA 31), which relies on case study data. The indicators "water abstraction rates' (IRENA 22), 'population trends of farmland birds' (IRENA 28) and 'share of agriculture in water use' (IRENA 34.3) are based on regular data sources. However, the data on water resources collected by the joint OECD/Eurostat questionnaire are not updated annually. The data on 'area under agri-environmental support' come from a recent data source (the Common indicators for monitoring rural development programmes were set up in 2000) that is annually updated.

4.8.6 Ease of interpretation

Only the water use (intensity) indicator (IRENA 10) provides messages that may be assessed as very clear to understand, even though it uses proxy information. The information from the population trends of farmland birds (IRENA 28) is clear, but attributing any changes in this indicator to agricultural water use is not possible at this stage.

4.8.7 Cost effectiveness

All but one of the indicators are based on existing statistics and data sets, which are also considered easily accessible (with the exception of IRENA 28, underpinned by a data source from BirdLife International — a non-government organisation). The indicator 'groundwater levels' (IRENA 31) has been compiled as a case study, using data related to one region of Spain.

Agricultural water use

Table 4.3   Evaluation of indicators used to undertake the environmental assessment of agricultural water use

Indicator criteria	Sub-criteria	Scoring	Driving forces — Water use (intensity)	Pressures — Water abstraction	State/impact — Ground water levels	State/impact — Population trends of farmland birds	State/impact — Share of agriculture in water use	Responses — Area under agri-environment support	Responses — Regional levels of good farming practices
	IRENA indicator		10	22	31	28	34.3	1	2
Policy relevance	Is the indicator directly linked to Community policy targets, objectives or legislation?	0 = No 1 = Yes, indirectly 2 = Yes, directly	1	1	2	1	0	2	2
	Could the indicator provide information that is useful to policy action/decision?	0 = Not at all 1 = Fairly useful 2 = Very useful	2	1	2	1	1	2	2
Responsiveness	Is the indicator responsive to environmental, economic or political changes?	0 = Slow, delayed response 1 = Fast, immediate response	1	1	0	0	0	1	0
Analytical soundness	Is the indicator based on indirect (or modelled) or direct measurements of a state/trend?	0 = Indirect 1 = Modelled 2 = Direct	0	2	2	2	2	2	2
	Is the indicator based on low/medium/high quality statistics or data?	0 = Low quality statistics/data 1 = Medium quality statistics/data 2 = High quality statistics/data	2	0	0	1	0	1	1
	What are the causal links with other indicators within the DPSIR framework?	0 = Weak or no link 1 = Qualitative link 2 = Quantitative link	1	0	0	1	0	1	1
Data availability and measurability	Good geographical coverage?	0 = Only case studies 1 = EU-15 and national 2 = EU-15, national and regional	2	1	0	1	1	1	1
	Availability of time series	0 = No 1 = Occasional data source 2 = Regular data source	2	2	0	2	2	1	0
Ease of interpretation	Are the key messages clear and easy to understand?	0 = Not at all 1 = Fairly clear 2 = Very clear	2	0	0	1	0	1	1
Cost effectiveness	Based on existing statistics and data sets?	0 = No 1 = Yes	1	1	0	1	1	1	0
	Are the statistics or data needed for compilation easily accessible?	0 = No 1 = Yes, but requires lengthy processing 2 = Yes	2	2	0	0	2	1	0
Total score			**16**	**11**	**6**	**11**	**9**	**14**	**10**
Classification of indicators: 0 to 7 (*) = 'Low potential' 8 to 14 (**) = 'Potentially useful' 15 to 20 (***) = 'Useful'			***	**	*	**	**	**	**
Final classification of Indicators according to the following criteria: Policy relevance at least 2 points, Analytical soundness at least 4 points, Data availability at least 3 points			**	**	*	**	**	**	**

5 Agricultural input use and the state of water quality

5.1 Summary of main points

- At EU-15 level the gross nitrogen balance in 2000 was calculated to be 55 kg/ha, which is 16 % lower than the balance estimate in 1990, which was 65 kg/ha. In 2000 the gross nitrogen balance ranged from 37 kg/ha (Italy) to 226 kg/ha (the Netherlands). All Member States show a decline in estimates of the gross nitrogen balance (kg/ha) between 1990 and 2000, apart from Ireland and Spain (22 % and 47 % increase, respectively). The following Member States showed organic fertiliser application rates greater than the threshold of 170 kg/ha specified by the nitrates directive in 2000: the Netherlands (206 kg/ha) and Belgium (204 kg/ha). The general decline in nitrogen balance surpluses is due to a small decrease in nitrogen input rates (– 1.0 %) and a significant increase in nitrogen output rates (10 %).
- The calculation of regional of gross nitrogen balances would provide a much better insight into the likelihood of nutrient losses to water bodies, in combination with data on farm management practices as well as climatic and soil conditions. Such an indicator could not be developed in the timeframe of the IRENA operation, mainly due to the lack of important data at regional level (manure, fertiliser application, yield coefficients) and even at national level (particularly the uptake of nitrogen through fodder and pastures).
- Livestock densities at NUTS 2/3 level give a regionalised picture of likely agricultural nutrient pressure. Regional concentrations of livestock linked to intensive pig and dairy production are found in the west of Germany, the Netherlands, Belgium, Brittany, northwest and northeast Spain, the Italian Po valley, Denmark, the west of the United Kingdom and southern Ireland.
- Monitoring data on nitrate concentrations in groundwater bodies and rivers are available but representative only at EU-15 level or for groups of Member States. Ground water concentrations of nitrates have largely remained stable between 1993 and 2002, apart from an apparent decline in southern EU Member States. Nitrate concentrations at river stations have declined slightly between 1992 and 2001 in Denmark, Germany, Luxembourg, the Netherlands and the United Kingdom and remained stable at lower levels in Austria, Finland and Sweden. France is the only country that shows a slight increase.
- The average share of agriculture in total nitrogen loading to surface waters for nine Member States (Austria, Belgium, Denmark, Finland, France, Germany, Italy, the Netherlands and Sweden) is 58 %. No data could be found for the other EU-15 Member States.
- Key responses from agricultural policy to nutrient leaching include the introduction of codes of good farming practice (GFP), which are mainly based on the legislative provisions of the nitrates directive, and agri-environment schemes. All Member States include requirements on nutrient management in their codes of GFP. The majority of national agri-environment programmes include measures relevant for agricultural nutrient management but currently available information provides little detail on the priority of such issues in national agri-environment scheme design.
- Data for the indicators pesticide soil contamination (IRENA 20) and pesticides in water (IRENA 30) rely on a modelling approach or case study material, respectively. Modelling is an important approach for overcoming lack of direct measurements but requires good input data. Further work needs to be carried out to improve the quality of these indicators.

5.2 Introduction

Water quality is a major environmental and health concern in Europe. Agriculture can affect water quality through the leaching or run-off of nutrients and pesticides. Farming is considered the main source of diffuse nitrogen pollution in Europe (EEA, 2003a).

The overloading of seas, coastal waters, lakes and rivers with nutrients (nitrogen and phosphorus) can affect the environment. Whereas phosphorus causes eutrophication of fresh water ecosystems due to its eutrophying effect, nitrates are considered to be harmful to human health. Nitrate is also known to be a major cause of eutrophication of coastal waters. Eutrophication is a process that in extreme circumstances results in massive blooms of planktonic algae. Decomposing algae can lead to oxygen depletion in water causing the death of fish and other aquatic organisms. Increased nutrient content can also lead to changes in the natural vegetation of water bodies. Phosphorus, primarily in the form of phosphate, is not as soluble as nitrate and is primarily transported by sediment in run-off. Nitrogen, on the other hand usually leaches through soils to ground waters as nitrates or is emitted as nitrous oxide from mineral fertilisers or as ammonia from livestock manure.

This chapter uses the relevant indicators to investigate the link between agricultural trends in fertiliser and pesticide use and water quality.

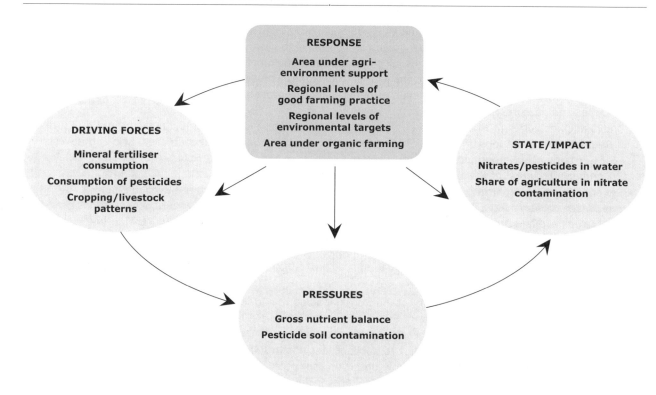

Figure 5.1 Environmental assessment of agricultural input use and state of water quality, based on the DPSIR framework

Table 5.1 IRENA indicators relevant for assessing agricultural input use and state of water quality

DPSIR	IRENA indicators	
Driving forces	No 8	Mineral fertiliser consumption
	No 9	Consumption of pesticides
	No 13	Cropping/livestock patterns
Pressures	No 18	Gross nitrogen balance
	No 20	Pesticide soil contamination
	No 21	Use of sewage sludge
State	No 30	Nitrates/pesticides in water
Impact	No 34.2	Share of agriculture in nitrate contamination
Responses	No 1	Area under agri-environment support
	No 2	Regional levels of good farming practice
	No 7	Area under organic farming

5.3 IRENA indicators related to agricultural input use and water quality

The Driving force — Pressure — State/impact — Response analytical framework provides a means to show linkages and associations between indicators (Figure 5.1 and Table 5.1), and to assess the relationship between agricultural fertiliser and pesticide use and state of water quality.

5.4 Agricultural driving forces

Agricultural driving forces related to pressures on water quality are trends in: mineral fertiliser consumption, including application rates (IRENA 8), the consumption of pesticides (IRENA 9) and cropping/livestock patterns (IRENA 13). Relevant trends include:

- Total nitrogen (N) mineral fertiliser consumption in EU-15 decreased by 12 % from 1990–2001. Total phosphate (P_2O_5) mineral fertiliser consumption in EU-15 decreased by 35 % from 1990–2001.
- The total estimated amount of pesticides used in agriculture increased by 20 % between 1992 and 1999.
- The number of livestock units of cattle and sheep decreased by 8.3 % and 3.4 %, respectively between 1990 and 2000 (EU-12). The livestock units of pigs, on the other hand, increased by 14.5 % between 1990 and 2000 (EU-12).

5.5 Agricultural pressures on water quality

Agricultural pressure indicators provide insight into the risks, which agricultural activities pose for water quality. Relevant indicators include: gross nitrogen balance (IRENA 18), soil pesticide contamination (IRENA 20), and use of sewage sludge (IRENA 21).

5.5.1 Gross nitrogen balance

Nutrient or mineral balances establish links between agricultural nutrient use, changes in environmental quality and the sustainable use of soil nutrients in terms of nutrient inputs and outputs. A persistent surplus indicates potential environmental problems; a persistent deficit indicates a potential risk of decline of soil nutrient status. However, as far as environmental impacts are concerned, the main determinant is the absolute size of the nutrient surplus/deficit linked to local farm nutrient management practices and agro-ecological conditions, which determine denitrification and absorption of nitrates in the soil.

As a general rule, data on inputs are estimated to be more accurate and reliable than data on outputs, because there is particular uncertainty about yield data for fodder and grass. As this uncertainty is carried through to the total N-balance, the same precautions should also be taken before drawing conclusions from the results of the total balance. There are also uncertainties in relation to the agronomic coefficients used, especially where there are large differences in farming conditions within a country.

At EU-15 level the gross nitrogen balance in 2000 was calculated to be 55 kg/ha, which is 16 % lower than the balance estimate in 1990, which was 65 kg/ha. In 2000 the gross nitrogen balance ranged from 37 kg/ha (Italy) to 226 kg/ha (the Netherlands). All Member States show a decline in estimates of the gross nitrogen balance (kg/ha) between 1990 and 2000, apart from Ireland and Spain (22 and 47 % increase, respectively). The following Member States showed average organic fertiliser application rates greater than the threshold of 170 kg/ha specified by the nitrates directive in 2000: Belgium (204 kg/ha) and the Netherlands (206 kg/ha). The general decline in nitrogen balance surpluses is due to a small decrease in nitrogen input rates (– 1.0 %) and a significant increase in nitrogen output rates (10 %).

For those Member States that had supplied confirmed national nitrogen balances to the OECD at the time of writing data was provided by the OECD secretariat. For the other EU-15 Member States the EEA calculated gross nitrogen balances according to the OECD/Eurostat methodology (OECD/Eurostat, 2003) recurring to Farm Structure Survey data for 1990 and 2000.

The breakdown of the nitrogen balance into its major input and output components in 2000 shows that the largest difference between Member States is the nitrogen originating from the net nitrogen input of manure (Figure 5.3). The application rates of organic fertilisers ranged from 31 N kg/ha in Spain to 206 N kg/ha in the Netherlands. However, the application rates of mineral fertilisers ranged from 35 N kg/ha in Austria to 179 N kg/ha in the Netherlands. 'Other nitrogen inputs' includes atmospheric deposition, biological nitrogen fixation, and seeds and planting material. This component is not as important as livestock manure or mineral fertilisers ranging from 8 N kg/ha in Portugal to 44 N kg/ha in the Netherlands.

Agricultural input use and the state of water quality

Figure 5.2 National gross nitrogen balances in 1990 and 2000 ([37]) ([38])

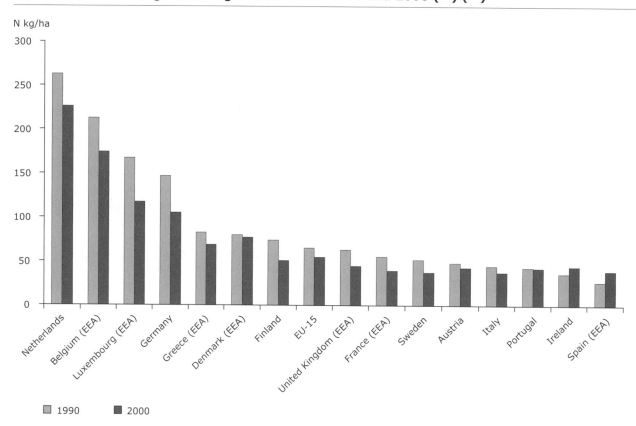

Source: OECD website (http://webdomino1.oecd.org/comnet/agr/aeiquest.nsf) and EEA calculations.

Figure 5.3 shows that the highest risk of nutrient leaching from agriculture is in the Netherlands, Belgium, Luxembourg and Germany, in spite of decreases in their gross nitrogen balance from 1990 to 2000. However, nitrogen leaching from soil to groundwater and rivers depends not only on the gross nitrogen balance but is significantly influenced by soil, climate and farm management.

National balances can mask important regional differences in the gross nutrient balances that determine actual nutrient leaching risk at regional or local level. Individual Member States can thus have acceptable gross nutrient balances overall at national level but still experience significant nutrient leaching in certain regions, for example in areas with high livestock concentrations. The calculation of regional gross nitrogen balances would provide a much better insight into the likelihood of nutrient losses to water bodies, in combination with data on farm management practices as well as climatic and soil conditions. Such an indicator could not be developed in the timeframe of the IRENA project, mainly due to the lack of important data at regional level (manure, fertiliser application, yield coefficients) and even at national level (particularly the uptake of nitrogen through fodder and pastures).

A map of overall livestock densities at NUTS 2/3 level is presented instead to give a regionalised picture of likely agricultural nutrient pressure (Figure 5.4). This shows regional concentrations of livestock linked to intensive pig and dairy production in the west of Germany, the Netherlands, Belgium, Brittany, northwest and northeast Spain, the Italian Po valley, Denmark, the west of the United Kingdom and southern Ireland.

5.5.2 Use of sewage sludge

The indicator focuses on the use of sewage sludge in agriculture as sufficient monitoring data on heavy metal or organic pollution in water is not available. It relates therefore less to 'water contamination' (which was the original concept of COM (144) 2001), than to the recycling of waste in agriculture.

([37]) In Belgium (Flanders) the first calculation is for 1998: in Sweden and Portugal the first calculations are for 1995.

Agricultural input use and the state of water quality

Figure 5.3 National nitrogen balances for 2000 split into major input and output components ([38])

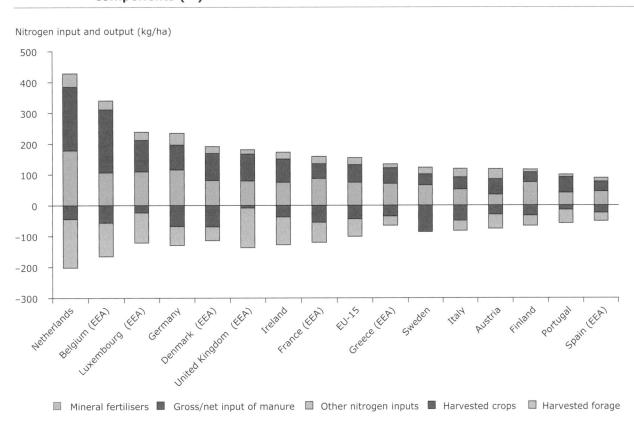

Source: OECD website (http://webdomino1.oecd.org/comnet/agr/aeiquest.nsf) and EEA calculations.

However, sewage sludge contains heavy metal concentrations that need to be monitored carefully. The indicator builds on data on volumes and heavy metal concentrations of sewage sludge that are submitted by Member States to the European Commission in the context of the requirements under the Standardised Reporting Directive (91/692/EEC). The Council Directive on the protection of the environment, and in particular of the soil, when sewage sludge is used in agriculture (86/278/EEC) lays down limit values for concentrations of heavy metals in the soil, in sludge and for the maximum annual quantities of heavy metals which may be provided to the soil. This indicator is therefore mainly covered in the following chapter.

5.5.3 Pesticide soil contamination

The pesticide soil contamination indicator (IRENA 20) uses a model to calculate the potential annual average content of herbicides in soils on the basis of computed pesticide degradation, which are a function of herbicide degradation properties and average monthly temperatures. The model takes into account the five most used herbicides per region. Time series of calculated potential annual average content of herbicides present in soils are analysed to detect potential trends under cereal, maize and sugar beet cultivation. The calculations indicate that ten of the EU-15 Member States face a statistically significant increasing trend for the modelled average quantity of herbicides present in soils under cereal cultivation. The calculations also indicated that Austria, France, Germany, Portugal and Spain face a statistically significant increasing trend for the modelled potential annual average content of herbicides present in soils under maize cultivation. In addition, Belgium, Italy, Luxembourg and Spain could face a statistically significant increasing trend for the modelled potential annual average content of herbicides present in the soils under sugar beet cultivation. On the other hand a statistically significant decreasing trend is calculated for sugar beet cultivation for Denmark.

([38]) The country name followed by (EEA) indicates balances that have been calculated by the EEA on the basis of EU level data sets.

Agricultural input use and the state of water quality

Figure 5.4 Regional distribution of cattle, sheep and pig livestock units (LU) per ha of UAA in 2000 and change from 1990–2000

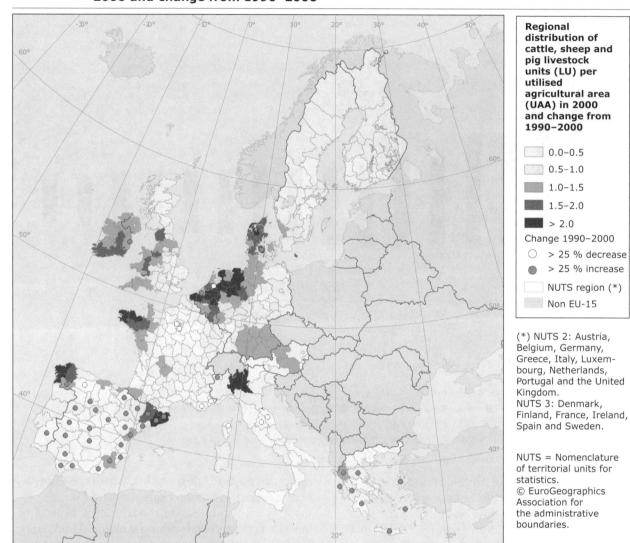

Note: Poultry figures are part of the calculation of national gross nitrogen balances but not included in this graph. Adding poultry production would emphasize some regional livestock hot spots, for example in the Benelux region.

Source: Community survey on the structure of agricultural holdings (FSS), Eurostat.

An increase of pesticide residues in the soil could also affect water quality through leaching into groundwater bodies or soil erosion processes. However, the information currently available is not sufficient to provide definite conclusions on trends in average annual pesticide content in soils, and even less so on water pollution risks. This highlights that further research and data collection are urgently needed in this area (see also Section 3.4.1.2).

5.6 State of/impacts on water quality

State of/impacts on water quality are shown by the indicators on nitrates and pesticides in water (IRENA 30), and the share of agriculture in nitrate contamination (IRENA 34.2).

5.6.1 Nitrates in water

Information on concentrations of nitrates in water is extracted from the Eurowaternet database maintained by the European Topic Centre on Water. IRENA 30 gives an overview of trends in nitrate concentrations in groundwater bodies and rivers across the EU-15, between 1990 and 2000.

5.6.1.1 Groundwater

The information on nitrate concentrations is based on 289 ground water bodies from 14 EU Member States. Care has to be taken in interpreting the data as the values are based on a limited number of samples as for example Belgium is only

Figure 5.5 Annual trends in the concentrations of nitrates (mg/l) monitored in groundwater (1993 to 2002) [39]

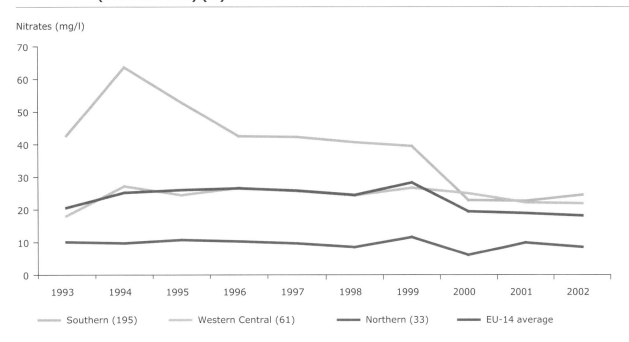

Source: EEA data service, 2004.

represented by one groundwater body, whereas France is represented by 74 groundwater bodies. For this reason Member States are grouped into three regions: southern Europe, western central Europe and northern Europe (Figure 5.6) [40].

In southern Europe, nitrate concentrations have declined since 1993 from around 40 to 25 mg NO_3/l. In western central Europe the concentrations have remained just above 20 mg/l NO_3, whereas in northern Europe nitrate concentrations have remained close to 10 mg NO_3/l. The EU-14 trend is strongly influenced by the changes in southern Europe. Generally, it can be said that concentrations have remained stable during the 1990s, varying between 20 and 30 mg NO_3/l. An important point to bear in mind, however, is that there is a delay of nitrate transfer from soil to groundwater depending on the soil type and geology (2–3 years for shallow waters in sandy soils, 10–40 years for deep waters in chalk limestone). The EU nitrates directive specifies a maximum concentration of 50 mg NO_3/l.

5.6.1.2 Rivers

Annual trends in the concentrations of nitrates (mg/l) monitored in rivers are available for eight EU-15 Member States (Figure 5.7). These Member States are grouped on the basis of data similarity:

- Denmark, Germany, France, Luxembourg, the Netherlands and the United Kingdom;
- Finland and Sweden;
- Austria.

Nitrate concentrations at river stations have declined slightly between 1992 and 2001 in Denmark, Germany, Luxembourg, the Netherlands and the United Kingdom and remained stable at lower levels in Austria, Finland and Sweden. Only France shows a slight increase in reported nitrate concentrations but remains close to the EU-15 average nitrate concentration levels.

5.6.2 Pesticides in water

The monitoring of pesticides is a challenging task due to the high number of registered pesticide

---

[39] The numbers in brackets indicate the number of groundwater bodies.
[40] The northern group comprises Denmark, Finland and Sweden. The western central group comprises Austria, Belgium, Germany, Ireland, the Netherlands and the United Kingdom. The southern group comprises France, Greece, Italy, Portugal and Spain.

Agricultural input use and the state of water quality

Figure 5.6   Annual trends in nitrate concentrations (mg/l) monitored in rivers (1992 to 2001)

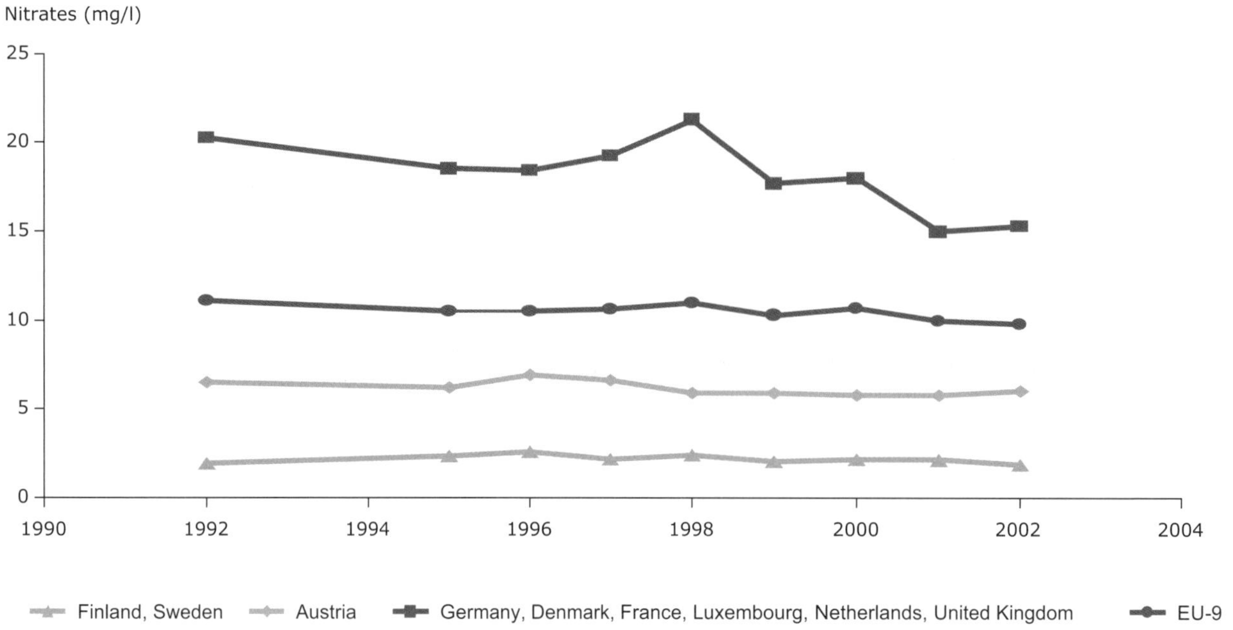

Nitrates (mg/l)

— Finland, Sweden   — Austria   — Germany, Denmark, France, Luxembourg, Netherlands, United Kingdom   — EU-9

Source: EEA data service, 2004.

substances. There is limited information available and a lack of reliable data on pesticides in ground and surface water. However, pesticide pollution is reported in a number of national reports. At present it is not possible to provide an EU-wide presentation and overview — instead a few national case studies are presented. Eurowaternet and other data show that there was a reduction of atrazine concentrations in groundwater in Austria, Belgium, Germany, England and Wales during the ten years to 2002 (but not in Denmark). In Austria at least this is likely to be the effect of the ban on the use of atrazine (UBA, 2001).

The Environment Agency of England and Wales (2003) reports a 23 % reduction in the share of fresh water samples with a pesticide concentration over 0.1 µg/l in 2003, compared with the mean for 1998–2002 (when an average 8–10 % of the samples exceeded the over 0.1 µg/l threshold for the pesticides measured)..

Monitoring of pesticide concentrations in France at 624 measurement points on rivers was used to classify water samples according to a grading system based on ecological and human health criteria. In the year 2002, 51 % were classified as good to high quality, 38 % as fair or poor quality and 8 %, were classified as very poor quality ([41]). Regarding water used for drinking water production, on 838 surface water catchments, 39 % required a specific treatment due to pesticide contamination and 1 % was unsuitable for drinking water production. For groundwater, 2603 catchments were monitored and the proportion was 21 % of the catchments needing specific treatment (IFEN, 2004).

Denmark reported the presence of pesticides in 37 % of the groundwater sources used for drinking water production, with 4 % having concentrations exceeding the drinking water standards (0.1 µg/l) (GEUS, 2004).

5.6.3   Share of agriculture in nitrate contamination

The share of agriculture in nitrate contamination is reported by some Member States in response to the OECD questionnaire underpinning the forthcoming report on Environmental Indicators for Agriculture Volume 4. The questionnaire requests information concerning the contamination of surface, ground and coastal waters. There are no responses regarding contamination of groundwater and only 2 out

---

([41])   3 % of the samples were not classified due to the absence of detectable pesticide concentrations.

Figure 5.7 Estimated share of agriculture in total nitrogen leaching to surface waters in 1995

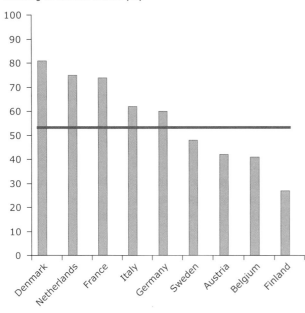

Note: The dark green line represents the average share of agriculture in nitrate contamination across nine EU Member States. To calculate this average Member State shares were weighted by their respective land area.

Source: OECD (2006, forthcoming) and UBA, 2001.

of 15 countries have replied to the question regarding coastal waters. There are eight responses for the agricultural share of nitrogen contamination in surface waters for the year 1995 and data were also found for Austria (Figure 5.7). For the nine EU-15 Member States that have provided data, the weighted average share of agriculture in nitrate contamination is 56 %. At the national level, the average ranges from 37 % in Finland to 81 % in Denmark. There is insufficient data for other years to analyse time series changes.

5.7 Responses

Environmental legislation is an important policy tool for protecting water quality. Various EU Directives address water issues, among them the 'water framework' and 'nitrates' directives. Policy measures that are important to the implementation of EU legislation are covered by the response indicators in this storyline: 'area under agri-environment support' (IRENA 1), and 'regional levels of good farming practice' (IRENA 2).

5.7.1 Area under agri-environment support

EU agri-environment measures allow Member States to grant support to farmers for a range of environmentally favourable measures, including better nutrient management, conversion to organic farming or extensification of livestock production (e.g. reduction of stocking densities). The information available at EU level only allows for a limited classification of agri-environment schemes by type of action. These types are: organic farming, input reduction measures (including integrated production), crop rotation, extensification, programmes concerning landscape and nature conservation, plant varieties under threat of genetic erosion and breeds in danger of extinction (Figure 5.8). To assess the extent to which the implementation of agri-environment measures is addressing agricultural nutrient management, the key environmental objectives behind individual agri-environment schemes, (soil conservation, water protection, and biodiversity preservation or landscape enhancement) need to be identified. This is not always possible on the basis of the information available.

Nevertheless, measures such as input reduction, crop rotation and extensification of farming can all be expected to have positive impacts on nutrient balances and nutrient management. Several agri-environment schemes classified in the 'other' category (see below), and the measures supporting organic farming are likely to improve agricultural nutrient management. In 2002, the most important type of agri-environment schemes in terms of area covered were those aimed at the reduction of inputs, which included integrated farming in most Member States (8.4 million ha), extensification of farming (2.4 million ha) and crop rotation (0.6 million ha). Together these covered 11.4 million ha and represented 40 % of the total agri-environment area across the EU-15. Organic farming conversion and maintenance contracts (2 million of ha) represented 7 % of total agri-environment area. Nearly 22 % of the agri-environment area was classified in the category 'other', which in some Member States includes horizontal measures covering organic farming and other environmental issues, among others also manure management. For example, the Rural Environment Protection Scheme (REPS), the Irish agri-environment programme, requires the elaboration of a farm nutrient management plan for the total area of the farm.

5.7.2 Regional levels of good farming practice

Codes of good farming practice (GFP) are a key policy response to encourage the promotion of better management practices, including nutrient

Agricultural input use and the state of water quality

Figure 5.8 Breakdown of area under agri-environment measures by type of action (2002)

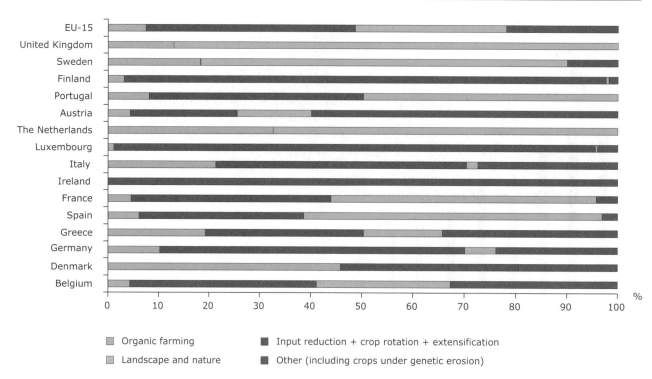

Note: The data only includes the area under agri-environment contracts signed in 2000, 2001 and 2002 under Regulation (EC) 1257/1999 (equivalent to a total of 22.7 million hectares). The schemes under the predecessor Regulation (EC) 2078/1992 (equivalent to a total of 11.3 million hectares) are not included, as only the total area and the area under organic farming is available. In Spain, the 'landscape and nature' category seems to include soil protection schemes and those aiming at the reduction of use of water for irrigation.

Source: DG Agriculture and Rural Development, Common indicators for monitoring the implementation of rural development programmes 2002.

management. The codes of GFP in combination with other policy instruments (training and advice) can be useful tools to minimise potential negative environmental effects of agricultural activity on water quality. Member States have to define codes of good farming practice at national or regional level in their rural development programmes (RDPs).

Numerous statutory and non-statutory standards may compose the codes of GFP. GFP usually entails, for example, compliance with the legislative provisions on the use of fertilisers (the 'nitrates directive'), pesticides, water management, waste, etc. Depending on the Member State or region concerned, farmers may also be required to follow guidelines of good farming practices concerning cropping patterns, soil conservation, pasture management and other farm management issues.

Most Member States have defined standards for fertilisation and handling of plant protection products, which are regulated at EU level (through the 'nitrates' and the 'Plant protection products' directives). However, there is a clear emphasis on these two aspects in Belgium (both regions), the Netherlands, Luxembourg, Austria, Denmark, Germany, Finland and the Italian region Emilia-Romagna. Some of these Member States have designated their whole territory (or an important part of it) as nitrate vulnerable zones (NVZs) and they have therefore defined compulsory requirements in the framework of their nitrate actions plans. These requirements are always included in the code of GFP and compliance with them is therefore a baseline condition for receiving payments under agri-environment schemes and Less Favoured Area allowances.

The management of fertilisers can also be considered as a priority issue in the codes of France, Ireland, Portugal and the United Kingdom, for those regions designed as NVZ. The United Kingdom, Sweden, the Wallonia region of Belgium and Portugal have set up some fertilisation standards for farms outside the NVZs (e.g. recommended fertilisation rates, restrictions on the timing of organic fertilisation, storage capacity), which are either recommendations or verifiable standards.

Table 5.2 Degree of coverage of farming practices relevant for nutrient management by national codes of GFP

Farming practices	BE-FI	BE-Wa	DK	DE	GR	ES	FR	IE	IT-ER	LU	NL	AT	PT	FI	SE	UK
Fertilisation	■	■	■	■	□	□	■	■	■	■	■	■	■	■	■	■
Pesticides	■	■	■	■	□	■	□	□	■	■	■	■	□	■	■	■
Waste management	—	□	□	—	□	□	□	□	□	□	□	—	□	□	—	□

■ Priority issue   — Issue not covered   □ Issue addressed

Source: Compiled from codes of GFP described in national rural development programmes 2000–2006.

Although all national codes include requirements in relation to the use of plant protection products, these are particularly detailed and strict in Germany (e.g. the legislation regulates the approval, accreditation for use, application and control of the spraying equipment, and lists further principles for pesticide use, which for the most part are not legally binding) and Ireland (with a 'statutory code of good plant protection practice').

Waste management includes the treatment of wastewater, prune residues, medicines, oils, packages, and other issues. Twelve Member States, address the issue in their codes as it can be important for the protection of water quality. For instance, in the region of Emilia-Romagna an administrative authorization is required as well as compliance with criteria specified in the legislation regarding disposal of wastewaters. In Greece, Spain and Denmark waste products have to be kept (stored) in accordance with the national regulation. In Finland unused pesticides and pesticide packages have to be destroyed and in the United Kingdom there are rules for sheep dip use and disposal in the groundwater protection code of agricultural practice (on an advisory basis).

5.7.3 Area under organic farming

Support for organic farming through payments under agri-environment schemes is a key response at EU and Member State level for promoting farming approaches that minimise the impact of agriculture on the environment. Also, the Commission has recently adopted a European action plan for organic food and farming, which is matched by national action plans in many Member States (see IRENA 3).

European studies have reported that organic cropping practices reduced nitrate leaching up to 50 % compared to conventional practices. Organic cropping systems control nitrate leaching by stabilising nitrogen in crop plants used in rotations. Adding organic matter to the soil stimulates the growth and reproduction of soil organisms, which also retain soil nitrogen in a relatively stable form (Stolze *et al.*, 2000).

Danish assessments derived from the mid-term evaluation of the water action plan II (Vandmiljøplan II) estimate that conversion during the action plan period reduced nitrogen leaching by an average of 33 kg N per ha compared with the average level of leaching from conventionally farmed areas in 1998. In particular the conversion from conventional to organic livestock farming had a significant effect whereas organic arable cropping can lead to nitrogen leaching at similar levels to conventional practices due to the higher use of manure for fertilisation than on conventional arable farms. (Jørgensen and Kristensen, 2003).

5.8 Conclusions: evaluation of indicators

5.8.1 Summary: general evaluation

The three indicators in this environmental storyline classed in the category 'useful' are the driving force indicators 'mineral fertiliser consumption' (IRENA 8) and 'cropping/livestock patterns' (IRENA 13) and the response indicator 'area under organic farming' (IRENA 7). The pressure indicator 'gross nitrogen balance' (IRENA 18) would also be very policy relevant if available at regional level. Despite the fact that 'mineral fertiliser consumption' is not provided at the regional level, the indicator receives a high score due to the other criteria. The other eight indicators are classed in the category 'potentially useful'. In most cases these indicators have not reached a level of development to be considered as useful because data availability and measurability and analytical soundness are inadequate. The information on the State/impact of pesticides is in particular difficult to obtain. None of the indicators are, however, regarded as of low potential.

Agricultural input use and the state of water quality

Table 5.3 Evaluation of IRENA indicators used to analyse agricultural use of fertilisers and pesticides and the state of water quality

Indicator criteria	Sub-criteria	Scoring	Driving forces			Pressures		State/impact		Responses			
			Mineral fertiliser consumption	Consumption of pesticides	Cropping/livestock patterns	Gross nitrogen balance	Pesticide soil contamination	Nitrates/pesticides in water	Share of agriculture in nitrate contamination	Area under agri-environment support	Regional levels of good farming practices	Regional levels of environmental targets	Area under organic farming
	IRENA indicator no		8	9	13	18	20	30	34.2	1	2	3	7
Policy relevance	Is the indicator directly linked to Community policy targets, objectives or legislation?	0 = No 1 = Yes, indirectly 2 = Yes, directly	1	2	1	2	2	2	1	2	2	2	2
	Does the indicator provide information that is useful to policy action/decision?	0 = Not at all 1 = Fairly useful 2 = Very useful	1	1	2	2	1	1	1	2	2	1	2
Responsiveness	Is the indicator responsive to	0 = Slow, delayed response 1 = Fast, immediate response	0	0	1	0	1	0	0	1	0	1	1
Analytical soundness	Is the indicator based on indirect (or modelled) or direct	0 = Indirect 1 = Modelled 2 = Direct	2	2	2	1	1	2	2	2	2	2	2
	Is the indicator based on low/medium/high quality statistics or data?	0 = Low quality statistics/data 1 = Medium quality statistics/data 2 = High quality statistics/data	1	1	2	1	1	1	0	1	1	0	2
	What are the causal links with other indicators within the DPSIR framework?	0 = Weak or no link 1 = Qualitative link 2 = Quantitative link	2	1	2	2	0	1	1	1	1	1	1
Data availability and measurability	Good geographical coverage?	0 = Only case studies 1 = EU-15 and national 2 = EU-15, national and regional	1	1	2	1	1	1	1	1	1	1	2
	Availability of time series	0 = No 1 = Occasional data source 2 = Regular data source	2	2	2	1	1	2	2	1	0	0	1

Indicator criteria	Sub-criteria	Scoring	Driving forces			Pressures		State/impact		Responses			
			Mineral fertiliser consumption	Consumption of pesticides	Cropping/livestock patterns	Gross nitrogen balance	Pesticide soil contamination	Nitrates/pesticides in water	Share of agriculture in nitrate contamination	Area under agri-environment support	Regional levels of good farming practices	Regional levels of environmental targets	Area under organic farming
	IRENA indicator no		8	9	13	18	20	30	34.2	1	2	3	7
Ease of interpretation	Are the key messages clear and easy to understand?	0 = Not at all 1 = Fairly clear 2 = Very clear	2	1	2	2	0	1	1	1	1	1	2
Cost effectiveness	Based on existing statistics and data sets?	0 = No 1 = Yes	1	1	1	1	1	1	1	1	0	1	1
	Are the statistics or data needed for compilation easily accessible?	0 = No 1 = Yes, but requires lengthy processing 2 = Yes	2	2	2	1	1	1	2	1	0	2	2
Total score			15	14	19	14	10	13	12	14	10	11	18
Classification of indicators: 0 to 7 (*) = 'Low potential' 8 to 14 (**) = 'Potentially useful' 15 to 20 (***) = 'Useful'			***	**	***	**	**	**	**	**	**	**	***
Final classification of Indicators according to the following criteria: Policy relevance at least 2 points, Analytical soundness at least 4 points, Data availability at least 3 points			***	**	***	**	**	**	**	**	**	**	***

The following sections present in more detail the evaluation of individual indicators according to the criteria set out in Section 2.3. Table 5.3 summarises the scoring for all indicators in this storyline.

5.8.2 Policy relevance

The indicators 'consumption of pesticides' (IRENA 9), 'gross nitrogen balance' (IRENA 18), 'use of sewage sludge' (IRENA 21) and 'nitrates/pesticides in water' (IRENA 30) and all the response indicators are directly linked to Community targets, objectives or legislation. However, 'cropping/livestock patterns' (IRENA 13) and 'gross nitrogen balance' are regarded as very useful for policy action/decision with data provided at regional level. The response indicators (IRENA 1 and 2) are also very useful as they show measures taken to improve water quality.

5.8.3 Responsiveness

Only the indicators 'area under agri-environment support' (IRENA 1) and 'cropping-livestock patterns' (IRENA 13) are regarded as being immediately responsive to environmental, economic or political changes.

5.8.4 Analytical soundness

All indicators are based on direct measurements, apart from 'gross nitrogen balance' (IRENA 18) and 'pesticide soil contamination' (IRENA 20), which are based on modelled estimates. In addition, 'Use of sewage sludge' (IRENA 21) is based on indirect data on the use of sewage sludge to derive potential risk of heavy metals contamination. The model to estimate gross nutrient balance is based on high quality data, with a strong quantitative link to environmental state and impact. However, the model to estimate pesticide soil contamination is based on medium-quality data (it uses statistics together with several coefficients), and is limited to the five most commonly used herbicides. There is therefore not a strong quantitative link to environmental state and impact.

The indicator 'Cropping/livestock patterns' (IRENA 13) scores maximum points in terms of analytical soundness for providing information relevant for monitoring nutrient levels. The most important indicators according to their link with the other indicators of the DSPIR framework are the indicators on 'mineral fertiliser consumption', 'cropping/livestock patterns', and 'gross nitrogen balance'.

5.8.5 Data availability and measurability

Regional and regular (long-term series) data is only provided by the 'cropping/livestock patterns' indicator (IRENA 13). The only other regional indicator is 'pesticide soil contamination' (IRENA 20), although the modelled estimates rely heavily on non-regional data (e.g. herbicide use). All the indicators are based on regular (long-term series) data, except for the indicators IRENA 1, 2, 3 and 'gross nitrogen balance' (IRENA 18).

5.8.6 Ease of interpretation

All indicators are clear or fairly clear to understand, apart from information on pesticide soil contamination, because it's modelled estimates are not sufficiently reliable to provide clear messages.

5.8.7 Cost effectiveness

All indicators are based on existing statistics and data sets, apart from 'regional levels of good farming practice', which is underpinned by qualitative information coming from rural development programmes for the period 2000–2006.

6 Agricultural land use, farm management (practices) and soils

6.1 Summary of main points

- Estimates based on the Pesera model indicate that the areas with the highest risk of soil erosion by water (i.e. more than 5 tonnes soil loss/ha/year) are located in southern and western Spain, northern Portugal, southern Greece and central Italy. No trend information is currently available.
- An estimated distribution of major classes of organic carbon in topsoil in the EU-15 shows that 45 % of agricultural area corresponds to soils with medium organic carbon content (good condition). Soils with low and very low organic carbon content account for about 45 %. Areas with low organic carbon content (0–1 %) appear mostly in southern Europe and correspond to areas with high soil erosion risk. No trend information is currently available.
- In 2000, approximately 56 % of the EU-15 arable land was covered 70 % of the year and 24 % of the arable land was covered 80 % of the year. Only 5 % and 4 % of the arable area were covered just 50 % and 40 % of the time throughout the year, respectively.
- In most Member States (e.g. Spain, Italy, Ireland, Luxembourg, Denmark, Greece, Portugal) conservation tillage involves less than 10 % of arable land. However, conservation tillage methods are increasingly being adopted in all the EU-15 Member States — most notably in Germany, Spain, Finland, France, Portugal and the United Kingdom.
- There are important land cover changes to and from forest/semi-natural and agricultural land in Spain, Portugal, and Italy.
- Agricultural policy responses in the area of soil protection include the introduction of codes of good farming practice (GFP) and agri-environment programmes. The majority of Member States include farming practices for soil management in their codes of GFP, with a clear focus on Portugal, Greece and Belgium. Many national agri-environment programmes include measures to protect soil from erosion and to improve soil conditions, although the information available at EU level does not allow identifying precisely the number of schemes targeted at soil quality.

6.2 Introduction

Soil is a natural resource that provides crucial agricultural and environmental functions (Blum and Varallyay, 2004). Soil sustains biological activity and productivity, regulates water and solute flow, and filters and buffers organic and inorganic materials. Soil quality is conserved or improved by land use decisions that consider these multiple functions, and can be impaired by decisions that focus solely on single functions, for example agricultural productivity. Some soil degradation processes can be linked to farm management decisions, such as increasing the intensity of tillage or the planting of crops that increase the amount of time the soil is left bare. Plants and plant residues can protect soils from erosion, reduce run-off of nutrients, increase organic matter, and enhance soil biodiversity.

This chapter uses the relevant IRENA indicators to show the effect of agricultural activities on soils. Agricultural land use and management practices are seen as key driving forces, which determine normal soil functions. In this respect, the focus lies on on-site soil issues, with estimation of soil erosion risk (IRENA 23) and soil quality (IRENA 29), rather than on off-site impacts (e.g. sediment transport).

6.3 IRENA indicators related to agricultural land use, farm management and soils

The Driving force — Pressure — State/impact — Response analytical framework provides a means to show linkages and associations between indicators (Figure 6.1 and Table 6.1), and to assess the relationship between agricultural land use, farm management and soils.

Agricultural land use, farm management (practices) and soils

Figure 6.1 Environmental assessment of agricultural land use, management and soils based on the DPSIR framework

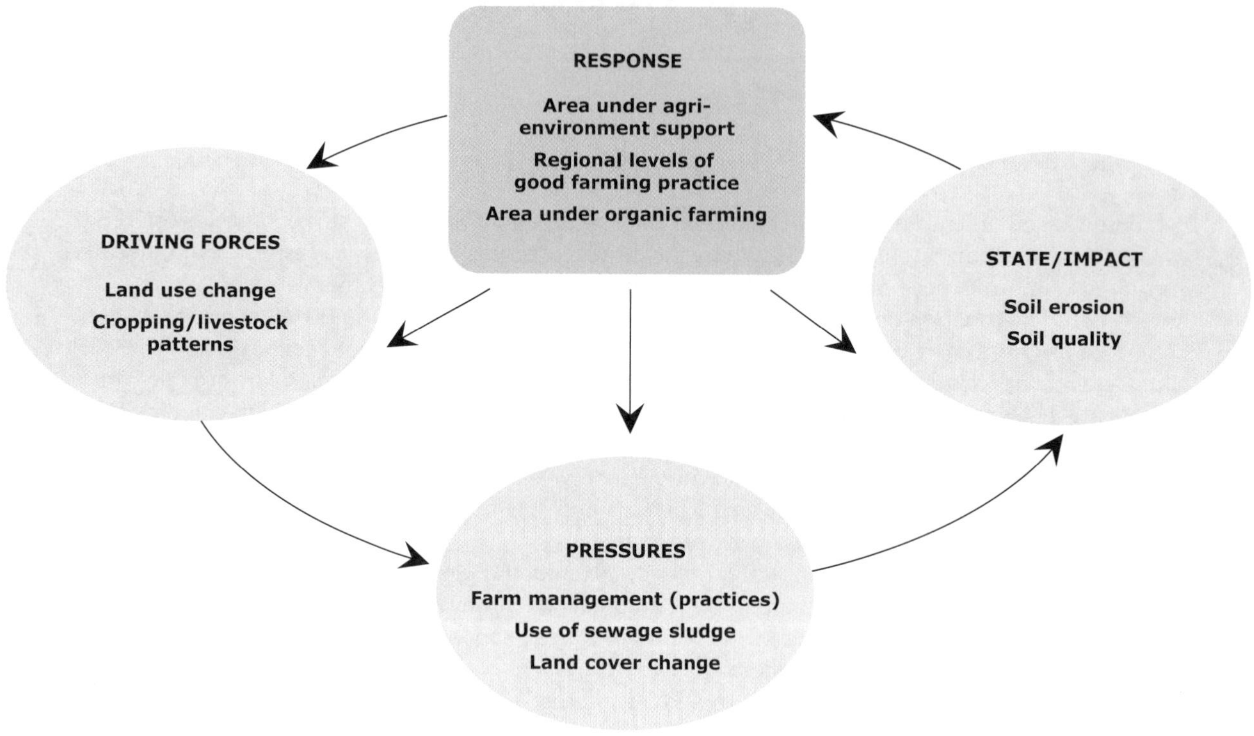

```
                    RESPONSE
                Area under agri-
                environment support
                Regional levels of
                good farming practice
                Area under organic farming

DRIVING FORCES                              STATE/IMPACT
Land use change                             Soil erosion
Cropping/livestock                          Soil quality
patterns

                    PRESSURES
                Farm management (practices)
                Use of sewage sludge
                Land cover change
```

Table 6.1 IRENA indicators relevant for assessing agricultural land use, management and soils

DPSIR	IRENA indicators	
Driving forces	No 12	Land use change
	No 13	Cropping/livestock patterns
Pressures	No 14*	Farm management practices
	No 21	Use of sewage sludge
	No 24	Land cover change
State	No 23*	Soil erosion
	No 29	Soil quality
Responses	No 1	Area under agri-environment support
	No 2	Regional levels of good farming practice
	No 7	Area under organic farming

* The indicator 23 is included in the state rather than the pressures domain as this corresponds better to its current development on the basis of the Pesera project. Indicator 14 is considered as pressure because agricultural cultivation methods and soil cover have a direct impact on the state of soil.

6.4 Agricultural driving forces

Agricultural driving forces related to the sustainable use of soil resources are: 'trends in cropping/livestock patterns' (IRENA 13), and 'land use change' (IRENA 12). Relevant key results from Chapter 3 on general trends in agriculture are:

- The trend of decreasing permanent grasslands and increasing arable land, which started in the 1970s, has continued during the 1990s. The largest decreases in permanent grasslands (more than 25 %) during the 1990s have occurred in Denmark and central and western France. Ploughing up permanent grasslands for arable agriculture increases the risk of soil erosion.
- During the period 1990 to 2000, the change in land use as from agriculture to artificial surfaces ranged from 2.9 % in the Netherlands to 0.3 % in France. In general the highest percentage of agricultural land (in 1990) converted to artificial areas (in 2000) is found close to major

Figure 6.2. Amount of sewage sludge used in agriculture in 1995 and 2000 (tonnes of dry matter)

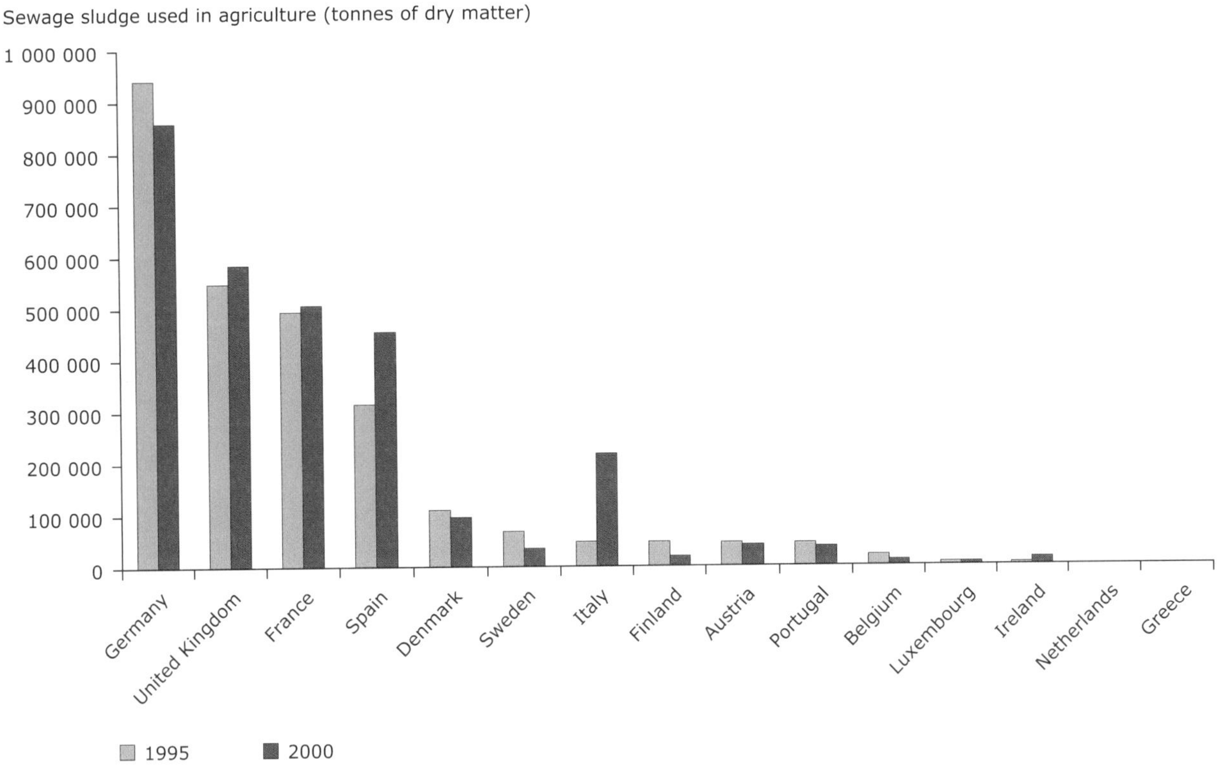

Sewage sludge used in agriculture (tonnes of dry matter); bars for 1995 and 2000 by country:
Germany ~940 000 / ~860 000; United Kingdom ~550 000 / ~585 000; France ~500 000 / ~510 000; Spain ~320 000 / ~460 000; Denmark ~120 000 / ~105 000; Sweden ~75 000 / ~45 000; Italy ~60 000 / ~230 000; Finland ~60 000 / ~30 000; Austria ~60 000 / ~55 000; Portugal ~60 000 / ~55 000; Belgium ~35 000 / ~25 000; Luxembourg ~20 000 / ~20 000; Ireland ~20 000 / ~30 000; Netherlands ~0 / ~0; Greece 0 / 0.

Source: Data submitted by Member States to the European Commission in the context of the requirements under the Standardised Reporting Directive (91/692/EEC).

conurbations. The conversion from agriculture to artificial surfaces results in soil sealing, which reduces the beneficial functions of soil.

6.5 Agricultural pressures on soil

Agricultural pressure indicators provide insight into the risks which agricultural activities are posing for the conservation of soil resources. Pressures include farm management practices (IRENA 14), in particular tillage practices and the management of soil cover, and land cover change (IRENA 24). Land cover change focuses on the land cover flows between agriculture and forest and 'semi-natural' areas, and between arable and pasture land.

6.5.1 Use of sewage sludge

The Council Directive on the protection of the environment, and in particular of the soil, when sewage sludge is used in agriculture (86/278/EEC) lays down limit values for concentrations of heavy metals in the soil, in sludge and for the maximum annual quantities of heavy metals which may be introduced into the soil. However, the use of sewage sludge should be carefully monitored for ensuring the non-accumulation of heavy metals in soil.

Between 1995 and 2000, the amount of sewage sludge used in agriculture increased in the United Kingdom, France, Spain, Italy and Ireland, whereas it decreased in Germany, Denmark, Sweden, Finland, Austria, Portugal and Belgium. The amounts of sewage sludge used in agriculture are negligible in the Netherlands and zero in Greece (Figure 6.2).

The reported concentrations of heavy metals detected in sewage sludge decreased for most metals between 1995 and 2000 (IRENA 21). All reported concentrations remained well within the standards set by Community legislation. The use of sewage sludge on farmland also has potential implications for water quality. However, given its importance for soil quality it is evaluated in this chapter.

6.5.2 Land cover change

IRENA 24 analyses the entries and exits to and from agricultural and forest/'semi-natural' land as

Agricultural land use, farm management (practices) and soils

Figure 6.3   Area of exits and entries from agriculture to natural/'semi-natural' land

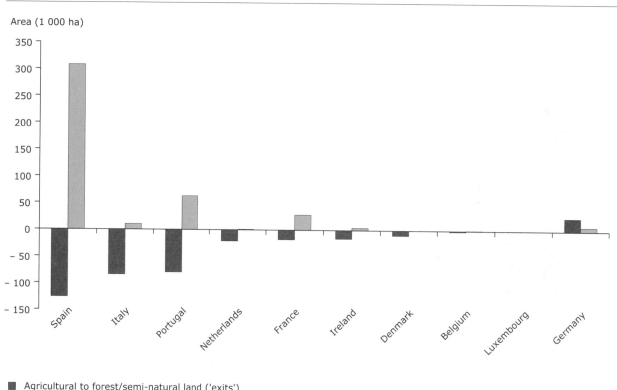

■ Agricultural to forest/semi-natural land ('exits')
□ Forest/semi-natural land to agricultural land ('entries')

Source: Corine land cover.

well as the land cover changes within agriculture (net arable and pasture area changes) between 1990 and 2000.

Among the Member States where Corine land cover data from 2000 are available ([42]) there are indications of a strong flow from forest/'semi-natural' land to agriculture land cover classes between 1990 and 2000 in Spain (300 000 ha). In Spain (126 000 ha), Italy (84 000 ha) and Portugal (80 000 ha) there are also indications of a strong flow from agriculture to forest/'semi-natural' land cover classes. In the other Member States these changes are much weaker (Figure 6.3).

It is worth noting that the analysed land cover flows occur in both directions, in Spain and Portugal in particular. This indicates that some of these changes relate to traditional rotation patterns in agro-forestry systems (intermittent arable cropping between grazing periods) rather than long-term conversion from one class to the other (see IRENA 24).

Comparisons with national land cover inventories (e.g. TERUTI in France) show that satellite-based observations may not have the same precision as survey-based approaches. Corine remains, however, the only data source that provides spatially referenced land cover information for all EU Member States.

6.5.3 Farm management

Soil cover and appropriate tillage practices are crucial for protecting soils from erosion and the decline of organic matter. Based on coefficients of the number of days in a year that a unit of arable land is covered by different crops, an estimation of the proportion of soil cover during the year 2000 has been carried out. In the year 2000, approximately 56 % of the EU-15 arable land was covered 70 % of the year and 24 % of the arable land was covered 80 % of the year. Only 5 % and 4 % of the arable area were covered just 50 % and 40 % of the time throughout the year, respectively. The regional map of soil cover by crop

---

([42])   At the time of report production, CLC 2000 was available for Belgium, Denmark, France, Germany, Ireland, Italy, Luxembourg, the Netherlands, and Portugal and Spain.

Figure 6.4 Regional map of soil cover (%) by crops 2000

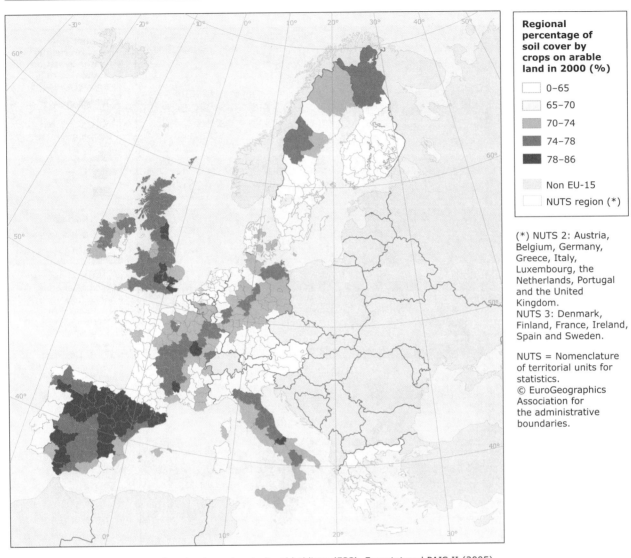

Source: Community survey on the structure of agricultural holdings (FSS), Eurostat and PAIS II (2005).

shows that in Italy, Denmark, Germany, northern Sweden and Portugal the soil is covered, on average, around 70 % of the year. In the United Kingdom and Spain soil cover is, on average, around 80 % of the year. The lowest degree of soil cover by crops is found in eastern Austria, Greece, south-western France, Finland and southern Sweden.

The duration of soil cover is directly influenced by the potential growing season (e.g. shorter growing season in Finland and Sweden than in France, predominance of spring crops in other Member States). However, soil erosion risk is increased when crops that are sown lengthen the period of bare soil or low soil cover during the year.

The adoption of conservation tillage methods (mulch tillage, minimum and reduced tillage) on arable land reduces some environmental impacts of agricultural activity on soil. There are few reliable statistics on adoption of conservation practices; available data come largely from associations for conservation agriculture and estimates from experts. In most Member States (e.g. Spain, Italy, Ireland, Luxembourg, Denmark, Greece, Portugal) conservation tillage involves less than 10 % of arable land. However, conservation tillage methods are being adopted increasingly in all the EU-15 Member States — most notably in Germany, Spain, Finland, France, Portugal and the United Kingdom.

6.6 State of soil

The state of soils is shown by the indicators on soil erosion (IRENA 23) and soil quality (IRENA 29).

Agricultural land use, farm management (practices) and soils

Figure 6.5 Annual soil erosion risk by water based on estimates of annual soil loss (aggregated results at NUTS 2/3 level)

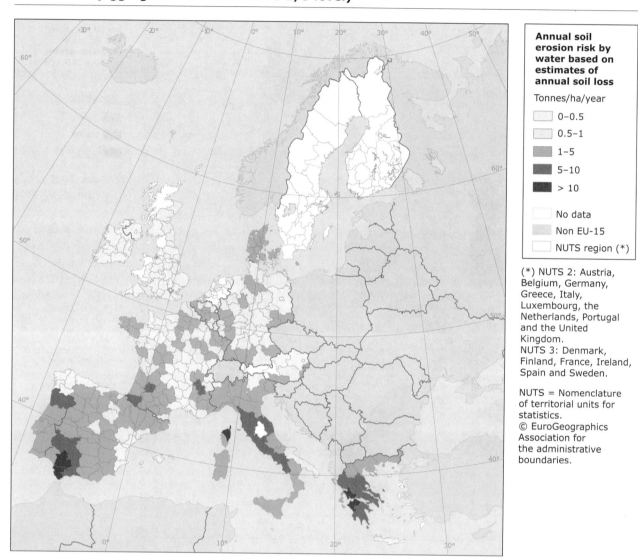

Source: Pesera project (Gobin and Govers, 2003).

6.6.1 Soil erosion

Soil erosion is a natural process that causes environmental concerns in situations of accelerated erosion, where the natural rate has been significantly increased by human activity (Gobin et al., 2004). The erosion rate at any given site is very sensitive to climate, topography and land use, as well as to particular soil conservation practices at farm level. The Mediterranean region is particularly prone to erosion because it is subject to long dry periods followed by heavy bursts of erosive rain falling on steep slopes with fragile soils. This contrasts with northern Europe where soil erosion is less serious because rain falls mainly on gentle slopes and is evenly distributed throughout the year. Consequently, the area affected by erosion is less extensive than in southern Europe (EEA, 2003b).

The Pan-European Soil Erosion Risk Assessment (Pesera) model uses a process-based and spatially distributed model to estimate soil erosion risk by water across Europe (Gobin and Govers, 2003). Two zones of erosion can be distinguished in EU-15 (Figure 6.5): a southern zone characterised by severe water erosion and a northern loess zone with moderate rates of water erosion. Within the two zones, there are areas where risk of erosion is more serious — the so-called hot-spots. Currently the Pesera model does not provide trend information.

The largest areas with a high erosion risk (i.e. a predicted loss of more than 5 tonnes/ha/year) lie in south-western Spain, northern Portugal, southern Greece and central Italy. The Pesera results do not always match with data and models employed at national level that can use more detailed information

Agricultural land use, farm management (practices) and soils

Figure 6.6 Estimated organic carbon content (%) in the surface horizon (0–30 cm) of soils in Europe

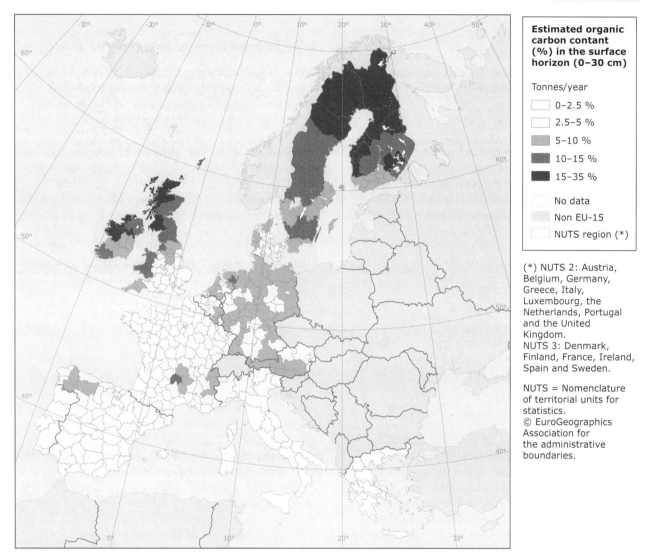

Source: Joint Research Centre, 2004.

but the model provides a good comparative overview at EU-15 level. Furthermore, Pesera focuses currently only on water erosion, not wind erosion.

6.6.2 Soil organic carbon

Currently, there is no consensus on the definition of soil quality. Nevertheless, the European Commission has listed environmental, economic, social and cultural functions of soil (COM 179 (2002)) for defining well functioning soils. Four out of five functions are directly linked to soil organic carbon content: food and biomass production, storage, filtering and transformation, habitats and gene pool and source of raw materials (*inter alia* peat). Soil organic carbon content is also very relevant in the context of the debate on climate change mitigation as soils can function as a potential sink for CO_2 that can be stored as organic carbon. Therefore, levels of soil organic carbon in topsoil have been adopted as a proxy indicator for soil quality since this covers both strictly agricultural criteria and wider environmental concerns.

Regional estimates of organic carbon content (%) in the top soil horizon are calculated using data on soil, land cover and temperature, with a model that estimates the rate of organic carbon degradation. At the moment time-series information is not available.

The distribution of organic carbon content across Europe (Figure 6.6) shows that areas of very low organic carbon content (0–1 %) appear mostly in southern Europe and correspond with areas with high soil erosion rates and warmer climates (see Figure 6.5). In northern Europe, highly organic soils (peat) are clearly highlighted. Agricultural suitability and carbon storage capacity do not increase beyond a level of 4–5 % organic matter content. In fact, peat soils can become a source of carbon dioxide and nitrous oxide emissions when they dry out and start to mineralise.

Humidity and temperature conditions strongly determine the content of organic carbon in soil, as decomposition of carbon is slowed by low temperatures and wet conditions. Therefore, soils in dry and warm Mediterranean areas are characterised by low to very low organic carbon content. This corresponds with areas that have been recognised in other indicator assessments as areas highly sensitive to desertification processes (DISMED, 2005).

6.7 Responses

The response indicators related to conserving soils are 'area under agri-environmental support' (IRENA 1), 'regional levels of good farming practice' (IRENA 2), and 'area of organic farming' (IRENA 7).

6.7.1 Area under agri-environment support

Under agri-environment measures EU Member States can grant support to farmers for environmentally favourable farm management, including soil conservation measures, conversion to organic farming or planting of landscape elements, such as hedgerows. The information currently available allows only for a limited classification of agri-environment schemes by type of action. These are: organic farming; a category including input reduction; crop rotation; extensification; landscape and nature conservation measures; and other measures including the preservation of rare plant varieties.

The measures for organic farming, input reduction, crop rotation and extensification of farming can be expected to favour soil conservation although their actual impact will vary considerably. In 2002, the most important type of agri-environment schemes in terms of area covered were those aimed at the reduction of inputs including in most Member States integrated farming, (8.4 million ha), extensification of farming (2.4 million ha) and crop rotation (0.6 million ha). Together these covered 11.4 million ha and represented 40 % of the total agri-environment area across the EU-15. The organic farming conversion and maintenance contracts (2 million ha) represented 7 % of the total agri-environment area (see Section 5.7.1).

These figures only refer to those schemes which started in the year 2000 or later and not to on-going commitments signed before.

6.7.2 Regional levels of good farming practice

Codes of good farming practice (GFP) are a key policy response to encourage the promotion of better management practices that will contribute to, amongst other objectives, improved soil management practices. The codes of GFP in combination with other policy instruments (training and advice) can be useful tools to minimise potential negative environmental effects of agricultural activity on soils. Member States have to define codes of good farming practice at national or regional level in their rural development programmes (RDPs). According to the IRENA assessment, all Member States, apart from France, the Netherlands, Sweden and Luxembourg, address soil management in their codes of GFP (Table 6.2).

Germany and Austria have quite detailed standards for soil protection, such as recommendations on soil coverage and crop rotation as well as compliance with regional and local regulations for soil protection and avoidance of erosion, respectively. In areas at risk in Austria, the local administration can set up measures such as minimum tillage or soil cover requirements, or make recommendations to minimise pressure on the soil. Sweden and Denmark have also addressed soil cover during autumn and winter in certain areas for avoiding water pollution from nitrate leaching.

In Portugal, Greece and in the Walloon region of Belgium, soil management can be considered as a priority issue. Many standards (mainly recommendations however) for soil management have been included in the codes of Portugal and Greece and, to a lesser extent, in Wallonia and the United Kingdom. Portugal, Greece and Belgium have a long list of recommendations relative to soil cover, crop rotation, cultivation practices (e.g. ploughing on slopes) and management of crop residues (e.g. the elimination of crop residues after harvest by burning is strictly forbidden and controlled), with more verifiable standards in Portugal and Greece than in Belgium. The codes of Spain and the United Kingdom have limited but verifiable standards relative to the prohibition of ploughing down the slope and the burning of crop residues and pastures. Verifiable standards on stocking density to avoid overgrazing are set out in Spain, Portugal, Greece and France. Practices related to pasture management seem to be a priority in the codes of France, Luxembourg, Ireland and the United Kingdom.

Table 6.2 Degree of coverage of farming practices relevant for soil protection by national codes of GFP

Farming practices	BE-Fl	BE-Wa	DK	DE	GR	ES	FR	IE	IT-ER	LU	NL	AT	PT	FI	SE	UK
Soil management	□	■	□	□	■	□	—	□	□	—	—	□	■	□	—	□
Pasture management	—	□	—	—	□	□	■	■	—	■	—	—	□	—	—	■

■ Priority issue   — Issue not covered   □ Issue addressed

Source: Compiled from codes of GFP described in national rural development programmes 2000–2006.

Most of the codes also include compliance with the requirements of the 'sewage sludge' Directive, targeted at the protection of soils when sludge is used in agriculture. There is particular focus on the use of sewage sludge in the codes of Portugal (e.g. prohibition of spreading near water sources for irrigation and domestic use), Austria (e.g. limits on amounts spread), Denmark (detailed rules concerning the period and areas of application of sludge, which must be part of a fertilisation plan).

6.7.3 Area under organic farming

By the end of 2002, the area devoted to organic farming (sum of organic and in-conversion area), certified under Regulation (EEC) No 2092/91, covered 4.8 million ha in EU-15. This represents an increase of 112 % over the period 1998–2002. Most assessments by researchers point to environmental benefits of organic farming for soil protection compared to conventional production. Both Stolze *et al.* (2000) and Shepherd *et al.* (2003) look at the impact of organic farming on soil properties. The general conclusion of these studies is that organic farming tends to conserve soil fertility and system stability better than conventional farming systems. Stolze *et al.* (2000) note that this is mostly due to higher organic matter contents and higher biological activity in organically farmed soils than in conventionally managed ones. Current data do not allow, however, (semi-)quantitative estimates of the likely benefits for soil conservation from the conversion to organic farming.

6.8 Conclusions: evaluation of indicators

6.8.1 Summary: general evaluation

Four indicators in this environmental storyline are classed in the category 'useful': the driving force indicators 'land use change' (IRENA 12), and 'cropping/livestock patterns' (IRENA 13), the pressure indicator 'land cover change' (IRENA 24), and the response indicator 'area under organic farming' (IRENA 7). The remaining seven indicators are classed in the category 'potentially useful', including 'soil erosion' (IRENA 23) and 'soil quality' (IRENA 29). This means that most of the pressure and all the state indicators have not reached a level of development to be considered as useful, mainly because of insufficient data availability and measurability and analytical soundness. Hence, to ensure comparable quality between the indicators, considerable improvements are needed. However, none of the indicators are regarded to have low potential.

The indicator 'farm management practices — tillage methods' (IRENA 14.1) has the lowest score. Information about new tillage practices (conservation agriculture) adopted by farmers is very relevant in relation to soil conservation, but few reliable data are available.

The following sections present in more detail the evaluation of individual indicators according to the criteria set out in Section 2.3. Table 6.2 summarises the scoring for all indicators in this storyline.

6.8.2 Policy relevance

The indicators on 'soil erosion' (IRENA 23) and 'soil quality' (IRENA 29), 'use of sewage sludge' (IRENA 21), and all the response indicators are directly linked to Community targets, objectives or legislation and therefore are considered to be fairly/very useful for policy decision/action. In their current version, however, the indicators 'soil erosion' and 'soil quality' do not incorporate reliable information on agricultural practices. In the absence of direct measurements of soil erosion or soil quality, the information provided by the indicators 'cropping/livestock patterns', 'management practices', 'soil cover' and 'land cover changes' is considered to be very important for policy making, as these are factors which influence soil erosion and organic matter

Agricultural land use, farm management (practices) and soils

Table 6.3 Evaluation of indicators used to undertake the environmental assessment of agricultural land use, management and soils

Indicator criteria	Sub-criteria	Scoring	Driving forces		Pressures				State		Responses		
			Land use change	Cropping/livestock patterns	Farm management (practices) — tillage	Farm management (practices) — soil cover	Use of sewage sludge	Land cover change	Soil erosion	Soil quality	Area under agri-environment support	Regional levels of good farming practices	Area under organic farming
		IRENA indicator no	12	13	14.1	14.2	21	24	23	29	1	2	7
Policy relevance	Is the indicator directly linked to Community policy targets, objectives or legislation?	0 = No 1 = Yes, indirectly 2 = Yes, directly	1	1	1	1	2	1	2	2	2	2	2
	Could the indicator provide information that is potentially useful to policy action/decision?	0 = Not at all 1 = Fairly useful 2 = Very useful	2	2	2	2	1	2	2	2	2	2	2
Respon-siveness	Is the indicator responsive to environmental, economic or political changes?	0 = Slow, delayed response 1 = Fast, immediate response	0	1	0	1	1	0	0	0	1	0	1
Analytical soundness	Is the indicator based on indirect (or modelled) or direct measurements of a state/trend?	0 = Indirect 1 = Modelled 2 = Direct	2	2	2	1	2	2	1	1	2	2	2
	Is the indicator based on low/medium/high quality statistics or data?	0 = Low quality statistics/data 1 = Medium quality statistics/data 2 = High quality statistics/data	2	2	0	2	1	2	1	1	1	1	2
	What are the causal links with other indicators within the DPSIR framework?	0 = Weak or no link 1 = Qualitative link 2 = Quantitative link	2	2	1	1	0	2	1	1	0	0	1
Data availability and measurability	Good geographical coverage?	0 = Only case studies 1 = EU-15 and national 2 = EU-15, national and regional	2	2	1	2	1	2	2	2	1	1	2
	Availability of time series	0 = No 1 = Occasional data source 2 = Regular data source	2	2	0	0	1	2	0	0	1	0	1

Agricultural land use, farm management (practices) and soils

Indicator criteria	Sub-criteria	Scoring	Driving forces		Pressures			State			Responses		
			Land use change	Cropping/livestock patterns	Farm management (practices) — tillage	Farm management (practices) — soil cover	Use of sewage sludge	Land cover change	Soil erosion	Soil quality	Area under agri-environment support	Regional levels of good farming practices	Area under organic farming
		IRENA indicator no	12	13	14.1	14.2	21	24	23	29	1	2	7
Ease of interpretation	Are the key messages clear and easy to understand?	0 = Not at all 1 = Fairly clear 2 = Very clear	2	2	1	2	1	1	2	2	1	1	2
Cost effectiveness	Based on existing statistics and data sets?	0 = No 1 = Yes	1	1	0	1	1	1	1	1	1	0	1
	Are the statistics or data needed for compilation easily accessible?	0 = No 1 = Yes, but requires lengthy processing 2 = Yes	1	2	1	1	1	1	1	1	1	0	2
Total score			17	19	8	14	12	16	13	13	13	9	18
Classification of indicators: 0 to 7 (*) = 'Low potential' 8 to 14 (**) = 'Potentially useful' 15 to 20 (***) = 'Useful'			***	***	**	**	**	***	**	**	**	**	***
Final classification: Policy relevance at least 2 points, Analytical soundness at least 4 points, Data availability at least 3 points.			***	***	**	**	**	***	**	**	**	**	***

content. 'Land use change' (IRENA 12) provides the only EU-wide information on soil sealing.

6.8.3 Responsiveness

The indicators evaluated as particularly sensitive to economic/political changes are 'cropping/livestock patterns' and 'farm management (soil cover)'. The response indicators related to the 'public policy' domain, particularly 'area under agri-environment support', and 'area under organic farming' are also able to capture changes. Indicator 2 ('regional levels of good farming practice') responds only slowly to external changes as it reflects medium to long term policy processes. 'Land use' and 'land cover change' indicators based on CLC are relatively slow to respond because any changes may not affect the dominant class and hence not be detected. The issues covered by the State/impact indicators are processes of soil change (degradation and loss of organic matter), which are relatively slow.

6.8.4 Analytical soundness

All indicators are based on direct measurements, apart from 'farm management practices' (soil cover) (IRENA 14.2), 'soil erosion' (IRENA 23) and 'soil quality' (IRENA 29), which are based on modelled estimates. The models to estimate soil erosion (IRENA 23) and soil quality-organic carbon content (IRENA 29) are based on medium quality data, which is collected in different time periods. Currently, trend information is not available.

The most important indicators according to their link with the other indicators of the DSPIR framework are the indicators on 'cropping/livestock patterns' and 'land cover change', components of which are used in the models estimating soil erosion and soil organic matter.

6.8.5 Data availability and measurability

Regional data are available for all but four indicators (IRENA 1, 2, 14.1 and 21,), which rely on national data. Time series data is available only for three indicators (12, 13, 24) and continuity is not yet fully assured for the Corine land cover based indicators. The indicators 'farm management practices' (tillage and soil cover) (IRENA 14.1 and 14.2), 'soil erosion' (IRENA 23), 'soil quality' (IRENA 29) and 'regional levels of GFP' (IRENA 2) do not currently offer trend information.

Agricultural land use, farm management (practices) and soils

6.8.6 Ease of interpretation

All indicators are clear to understand, apart from information on 'management practices — tillage', 'area under agri-environment support' and 'environmental targets' and 'use of sewage sludge', because the data are not sufficiently detailed or targeted to provide clear messages related to soil.

6.8.7 Cost effectiveness

All indicators are based on existing statistics and data sets, apart from 'farm management practices — tillage methods' (IRENA 14.1), which is based on results of the PAIS II project. The questionnaire employed in this project would need to be repeated, with a higher response rate to make the results reliable. The indicators reliant on modelled data require lengthy processing to access or include new data until the models are made operational. This is the same for the Corine land cover based indicators.

7 Climate change and air quality

7.1 Summary of main points

- Agriculture contributed around 10 % of total greenhouse gas emissions and about 94 % of ammonia emissions in the EU-15 in 2002.
- The greenhouse gases emitted by agriculture are nitrous oxide and methane, both of which have a far greater global warming potential than carbon dioxide. Agriculture also consumes fossil fuels for farm operations, thus emitting carbon dioxide.
- Emissions of greenhouse gases by the agriculture sector — methane and nitrous oxide — fell by 8.7 % between 1990 and 2002. This is due mainly to a 9.4 % reduction in methane from reduced livestock numbers and an 8.2 % reduction in nitrous oxide from decreased nitrogenous fertiliser use and changed farm management practices.
- Within the EU-15, emissions of ammonia to the atmosphere from agriculture decreased by 9 % between 1990 and 2002. The majority of this reduction is likely to derive from a reduction of livestock numbers across Europe (especially cattle), and the lower use of nitrogenous fertilisers across the EU-15.
- All EU-15 Member States have action plans for climate change and air quality. Most plans and programmes under the National Emission Ceilings (NEC) Directive include measures to reduce ammonia emissions from agriculture due to their negative health and environmental effects. However, only Ireland has a climate change action plan with specific targets for the agriculture sector. According to current projections (which exclude potential effects of the 2003 CAP reform) many Member States are likely to miss their 2010 ammonia reduction target under the NEC directive.
- The agriculture sector can make a positive contribution to reducing greenhouse gases through the production of bio-energy, thus substituting for fossil fuels. Agriculture at present contributes 3.6 % of total renewable energy produced and 0.3 % of total primary energy produced in the EU.
- With regard to CO_2, the role of agriculture in a climate change context is quite complex. Information derived from the IRENA indicators does not currently allow assessing the potential role of agricultural soils as CO_2 sink or, on the other hand, the large emissions of CO_2 from agricultural soils that are reported by some countries.

7.2 Introduction

At the international level, the EU is a key player in the effort to combat climate change. The EU was very active in the development of the two major treaties addressing the issue: the United Nations Framework Convention on Climate Change and the Kyoto Protocol. As a consequence of the 1997 Kyoto Protocol and the subsequent 1998 EU Burden Sharing agreement (Council Decision 2002/358/EC), the EU is committed to achieving an 8 % reduction in emissions of six greenhouse gases (GHG) by 2008–2012 compared to 1990 level. These include carbon dioxide (CO_2), methane (CH_4), nitrous oxide (N_2O), hydro fluorocarbons (HFCs), per fluorocarbons (PFCs), and sulphur hexafluoride (SF6). The European Commission has also acknowledged the requirement to extend emission reductions of greenhouse gases beyond the initial 2008–2012 Kyoto period. Recent EU presidency conclusions suggest emission reduction pathways for the group of developed countries in the order of 15–30 % by 2020 compared to the 1990 baseline envisaged in the Kyoto Protocol (European Council, March 2005).

Burning of fossil fuels, resulting in the emission of CO_2, is the main contributor to the greenhouse effect. The agriculture sector contributed 10.1 % of the total EU greenhouse gas emissions in 2002. The agriculture sector is largely responsible for emissions of two major greenhouse gases, methane and nitrous oxide. One tonne of methane is effectively 21 times more powerful as a greenhouse gas than one tonne of CO_2 in terms of its global warming potential.

Methane is emitted from three sources within livestock production systems: digestive processes

Climate change and air quality

in animals (enteric fermentation); anaerobic decomposition processes in animal manure; and anaerobic decomposition processes of waste products from animal processing. The latter often occur when large numbers of animals are managed in confined areas (e.g. dairy farms, beef feedlots, and pig and poultry farms). The production of methane is therefore closely related to livestock production, and to a certain extent, to the production type. About 1 % of methane emission also arose from rice cultivation in the period of 1990–2002 (IRENA 19).

Nitrous oxide is emitted during manure storage when the nitrogen in manure is converted to nitrous oxide, and by the conversion of nitrogen in the soil. These are natural processes, which are enhanced by agriculture. The sources of these emissions include synthetic fertilisers, animal waste, sewage sludge applications, biological nitrogen fixation and crop residues. One tonne of nitrous oxide is 310 times more powerful in terms of global warming potential than one tonne of CO_2. However, carbon dioxide is a minor agricultural greenhouse gas, contributing a share of only 0.06 % of total EU emissions of CO_2. The IRENA 7 indicator fact sheet presents a number of scientific references on the environmental impacts of organic agriculture compared to conventional agriculture. This literature review has shown that 'On a per-hectare scale, the CO_2 emissions are 40–60 % lower in organic farming systems than in conventional ones, whereas on a per-unit output scale, the CO_2 emissions tend to be higher in organic farming systems' (Stolze et al., 2000). Quantitative research information on N_2O emissions from manure and soil is scarce, but Stolze et al. (2000) conclude nevertheless that N_2O emissions per hectare on organic farms tend to be lower than on conventional farms, while N_2O emissions per kg of milk are equal or higher, respectively. Research results on methane emissions in different farming systems were reported by Shepherd et al. (2003).

Air pollution is one of the major issues mentioned in the 6th environment action programme (6EAP) and the focus of the Community programme 'Clean Air for Europe' (CAFE). The aim of the CAFE programme is to develop long-term, strategic and integrated policy advice to protect against significant negative effects of air pollution on human health and the environment. It is envisaged that the work from CAFE will lead to the adoption of a thematic strategy on air pollution under the sixth environmental action programme in 2005. Ammonia is one of the pollutants included under the CAFE programme as it contributes to environmental issues, such as acidification and eutrophication, and poses a potential risk to human health (McCubbin et al., 2002).

7.3 IRENA indicators linked to climate change and air quality

The Driving force — Pressure — State/impact — Response analytical framework provides a means of showing linkages and associations between indicators (Figure 7.1 and Table 7.1) and assessing the relationship between agriculture, climate change and air quality. Trends in national greenhouse gas emission levels (IRENA 19 and 34.1) and energy use by agriculture (IRENA 11) are used to assess the greenhouse gas emissions originating from agricultural activities. The link between agricultural energy use and emissions still needs to be explored with regard to deriving a better estimate of CO_2 emissions resulting from agricultural operations. As regards air quality, the ammonia indicator (IRENA 18sub) is used to estimate the quantity of NH_3 emissions from livestock. In addition, trends in the production of renewable energy sources (IRENA 27), such as wood from short rotation coppice and bio diesel, are described.

The key messages taken from Chapter 4 on general trends in agriculture in relation to climate change and air quality are:

- Motor fuels and lubricants are the main source of final energy consumption by agriculture in the EU-15, except for the Netherlands, which depends mainly on natural gas. Consumption of motor fuels and lubricants amounts to over half of total energy costs in most Member States.
- Energy is also used by agriculture in an indirect manner for the production of agrochemicals (e.g. fertilisers), farm machinery and buildings. Considerable amounts of natural gas are used for the production of inorganic nitrogen fertilisers. In the Netherlands, for example, three quarters of total amounts of natural gas used in the fertiliser industry is used for non-energetic purposes, serving as input for the production process.
- Total nitrogen (N) mineral fertiliser consumption in EU-15 has decreased by 12 % from 1990–2001.
- The number of livestock units of cattle decreased by 8.3 % between 1990 and 2000 (EU-12). The livestock units of sheep decreased by 3.4 % between 1990 and 2000 (EU-12). The livestock units of pigs, on the other hand, increased by 14.5 % between 1990 and 2000 (EU-12).

Figure 7.1   Environmental assessment of agriculture in relation to climate change and air quality based on the DPSIR framework

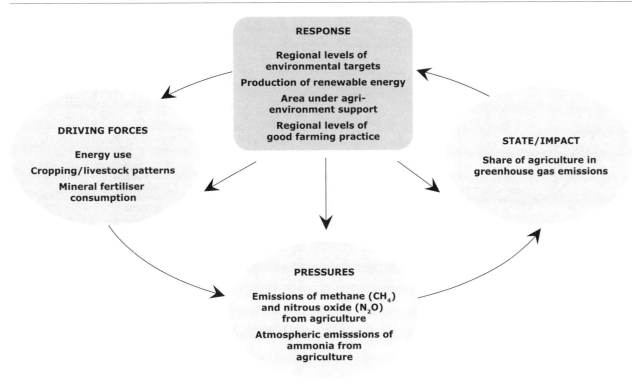

Table 7.1   IRENA indicators relevant to the assessment of agriculture in relation to climate change and air quality

DPSIR	IRENA indicators		Issue
Driving forces	No 8	Mineral fertiliser consumption	Climate change/air quality
	No 11	Energy use	Climate change
	No 13	Cropping/livestock patterns	Climate change/air quality
	No 14	Farm management practices — manure	Climate change/air quality
Pressures	No 18sub	Atmospheric emissions of ammonia from agriculture	Air quality
	No 19	Emissions of methane (CH_4) and nitrous oxide (N_2O) from agriculture	Climate change
Impact	No 34.1	Share of agriculture in greenhouse gas emissions	Climate change
Response	No 1	Area under agri-environment support	Climate change/air quality
	No 2	Regional levels of good farming practice	Climate change/air quality
	No 3	Regional levels of environmental targets	Climate change/air quality
	No 27	Production of renewable energy *	Climate change

*   This indicator has been placed in a different DPSIR category than proposed in COM (2001) 144. This arises from the need to present a logical storyline.

7.4   Agricultural pressures on climate change and air quality

Methane and nitrous oxide are the main sources of greenhouse gas emissions in the agricultural sector. Agricultural soils are a potential sink for CO_2 through carbon sequestration; however this is difficult to observe and measure (see Chapter 6, Section 6.6.2 on soil organic carbon).

7.4.1   Emissions of methane and nitrous oxide from agriculture

In absolute amounts, the agriculture sector emitted 416 million tonnes CO_2 equivalent of greenhouse gases in 2002. This is an 8.7 % reduction compared with 1990 emissions. Reductions in greenhouse gas emissions were mainly due to a 9.4 % reduction in methane enteric fermentation emissions because

Climate change and air quality

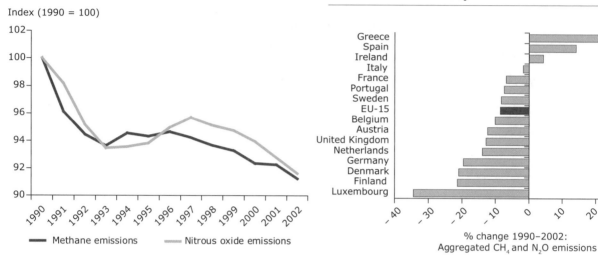

Figure 7.2a Methane and nitrous oxide emissions from agriculture 1990–2002 (EU-15 Member States) indexed relative to 1990 emission levels

Figure 7.2b Change in aggregated emissions of methane and nitrous oxide (ktonnes CO_2 equivalent) from agriculture 1990–2002 (EU-15 Member States)

Note: Emissions from agricultural transport and energy use are excluded, as these sectors are not defined as part of the agriculture sector by the current IPCC guidance.
Source: EEA, 2004a.

of a reduction in cattle numbers ([43]), and an 8.2 % reduction in nitrous oxide emissions from agricultural soils because of a decrease in the use of nitrogenous fertilisers (Figure 7.2). Although emissions fell on average, there are varying trends in different Member States. Luxembourg (34 % decrease), Finland (21 % decrease) perform better than the EU-15 average, while Greece (22 % increase) and Spain (14 % increase) perform worse.

7.4.2 Ammonia emissions

In Europe, ammonia emissions mainly occur as a result of volatilisation from livestock excretions, whether from livestock housing, manure and slurry storage, urine and dung deposition in grazed pastures or after manure spreading onto land. A smaller fraction of ammonia emissions results from the volatilisation of ammonia from nitrogenous fertilisers and from fertilised crops. Wherever possible, Member States are recommended to use country-specific emission factors that take into account the differences between Member States with respect to environmental and agricultural practice. Increasingly, Member States are performing research into developing, and subsequently using, regional-specific emission factors. However, research to develop and verify emission factors is both costly and time-consuming and so experimental data is lacking at both the national and European scale that would allow improved and more specific methods of calculating emissions to be developed (IRENA 18sub).

Ammonia, together with emissions of sulphur dioxide and nitrogen oxides, contributes to acidic deposition on soils and aquatic ecosystems, resulting in eutrophication. Excess levels of soil acidity affect the solubility of both essential and toxic elements, which can be particularly damaging on weakly buffered soils (clearly seen in woodlands). Furthermore, fine particulate matter pollution associated with ammonia can have serious negative consequences for human health (Amann et al., 2005; McCubbin et al., 2002).

In 2002, estimations indicate that the EU-15 agricultural sector emitted a total of 3 million tonnes of ammonia and was responsible for 94 % of total ammonia emissions in these countries. Within the EU-15, emissions of ammonia from agriculture have decreased by 9 % between 1990 and 2002. Most of this reduction is likely to derive from a reduction in livestock numbers

---

[43] In IRENA 19 'Emissions of methane (CH_4) and nitrous oxide (N_2O) from agriculture' cattle and other data related to agriculture are derived from FAO statistics to ensure international comparability of figures. This results in differences with livestock numbers used in IRENA 13 'Cropping/livestock patterns', which uses FSS data and expresses livestock numbers, in terms of livestock units.

Climate change and air quality

across Europe (especially cattle), and the lower use of nitrogenous fertilisers across the EU-15.

Ammonia emissions from agriculture, expressed in kilograms per utilised agricultural area, show a decrease in all EU-15 Member States,

Figure 7.3   Changes in ammonia emissions from agriculture (kg/ha) between 1990 and 2002

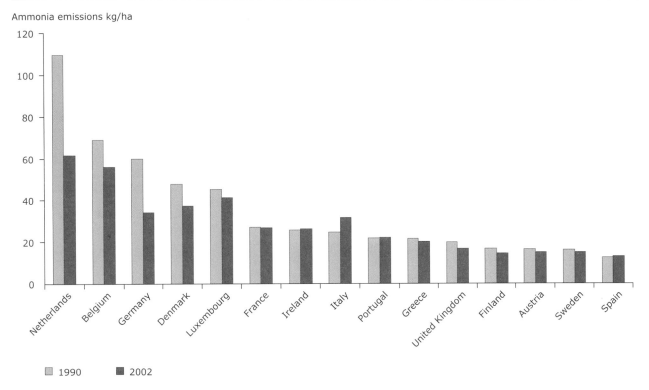

Source: Data reported by Member States to the UNECE/EMEP Convention on Long-range Transboundary Atmospheric Pollution (CLRTAP). Data on utilised agricultural area come from the Farm Structure Survey, Eurostat.

Figure 7.4a   Share of the EU-15 agricultural sector in total greenhouse gas emissions (2002)

Figure 7.4b   National share of the agriculture sector in total greenhouse gas emissions (2002)

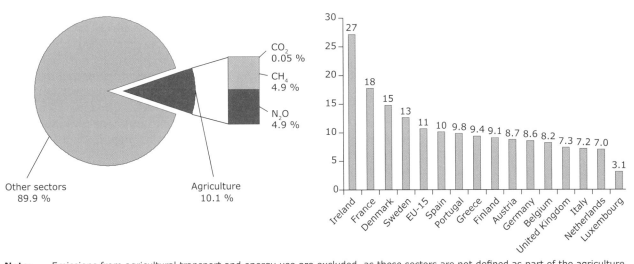

Note: Emissions from agricultural transport and energy use are excluded, as these sectors are not defined as part of the agriculture sector by the current IPCC guidance.

Source: EEA, 2004a.

apart from Italy, Spain, Ireland and Portugal, where (small) increases have occurred (Italy 28 %, Spain 6 %, Ireland 2 % and Portugal 1 % — Figure 7.3). Differences between Member States are due generally to the types and numbers of livestock together with other factors previously described, such as climatic conditions, agricultural management and storage installations. Cattle farming accounts for about 40 % of the EU-15 emissions, other livestock (mainly poultry and pigs) account for about 33 %, and the rest is coming from the use of nitrogen mineral fertilisers and other sources (Amann et al., 2005).

7.5 Impact on climate change and air quality

7.5.1 The share of agriculture in greenhouse gas emissions

Agriculture can act as a source as well as a sink for greenhouse gases although the potential for the latter is difficult to quantify. The sector is a major source of the non-CO_2 greenhouse gases methane and nitrous oxide. Both of these gases are many times more powerful greenhouse gases than CO_2. Agriculture contributed around 10.1 % of the total EU-15 emissions of greenhouse gases in 2002, whereas this share was 10.8 % in 1990. Ireland (27 %), France (18 %) and Denmark (15 %) had respective shares of agriculture emissions to total greenhouse gas emissions significantly higher than the EU average (Figure 7.4 a) and b)).

In the EU-15, greenhouse gases are emitted primarily from the combustion of fossil fuels for energy use in the energy production, industry, transport and residential sectors. Carbon dioxide (CO_2) emissions account for approximately 82 % of total greenhouse gas emissions in the EU and 95 % of these are energy-related. In contrast, the agriculture sector contributed around 10 % of the total EU greenhouse gas emissions in 2002, showing a small decrease since 1990 (i.e. less than 1 %).

7.6 Responses

7.6.1 Regional levels of environmental targets

The main factors which have influenced EU emissions of greenhouse gases from the agriculture sector since 1990 have been general underlying agro-economic trends. Certain policy measures under the common agricultural policy (CAP) and the implementation of the nitrates directive (EEA,

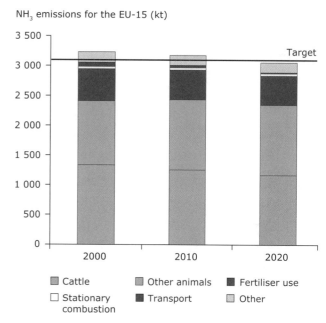

Figure 7.5 Projections of ammonia emissions to 2020 for the EU-15

Note: The NH_3 2010 emission ceiling under the NEC directive for the EU-15 is 3 110 kt.

Source: CAFE Programme scenario (Amman et al., 2005).

2004b) have also contributed to lower cattle numbers in individual Member States and to some changes in agricultural practices across the EU, such as a reduced use of nitrogenous fertilisers. All EU-15 Member States have action plans for climate change and air quality. Most plans and programmes under the National Emission Ceilings (NEC) Directive (2001/81/EC) include measures to reduce ammonia emissions from agriculture. However, only Ireland has a climate change action plan with specific targets for the agriculture sector.

The CAFE programme has carried out a substantial study on current and future trends in ammonia emission across EU Member States. According to these projection (which exclude potential effects of the 2003 CAP reform), many Member States are likely to miss their 2010 ammonia emission reduction target under the NEC directive. Figure 7.5 shows the EU-15 ammonia emissions projected by the CAFE programme against the EU-15 2010 target.

7.6.2 Regional levels of good farming practice

Codes of good farming practices correspond to the type of farming that a reasonable farmer would follow in the region concerned. This includes at the very least, compliance with general statutory environmental requirements, such as the 'nitrates' directive. Relevant standards regarding manure

Figure 7.6 Production of renewable energy from agricultural sources (EU-15)

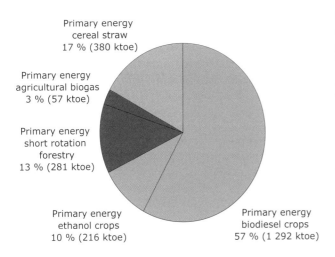

Source: European Bio diesel Board (2003), EuObserv'ER (2004), Eurostat FSS, Eurostat RES, Statistics Sweden, International Energy Agency, Faostat.

and slurry management and storage are therefore included in codes of GFP (see Table 5.2 concerning fertilisation and waste management). These codes are eligibility conditions for support under agri-environment programmes and the Less Favoured Area scheme. However, a substantial part of intensive livestock farms do not participate in these schemes and do not need to respect such codes, although they have to comply with the legal provisions of the 'nitrates' Directive.

7.6.3 Area under agri-environment support

Good farming practices also constitute the baseline requirements for farmers wishing to join agri-environment schemes. Agri-environment measures are designed to encourage farmers to protect and enhance the environment on their farmland. Farmers commit themselves, for a five-year minimum period, to adopt environmentally friendly farming techniques which go beyond the usual good farming practices. Some agri-environment schemes that have potential to impact positively on climate change include organic farming, input reduction and extensification schemes (see Section 5.7.1). Other actions taken at the national level to minimise ammonia reduction can include the phasing-out of slurry spreading onto the soil surface, which in Denmark has been replaced by direct soil incorporation of slurry into the soil. While this has resulted in a decrease in ammonia emission, certain side effects such as increased nitrous oxide emission and nitrate run-off have been identified (IRENA 18sub).

7.6.4 Production of renewable energy (by source)

The burning of fossil fuels is the main source of emissions of CO_2, which is responsible for climate change. Bio fuels — fuels produced from renewable sources — can be used as substitutes for fossil fuels. This can reduce the net emissions of greenhouse gases into the atmosphere as well as the associated emissions, if energy used during crop production and fuel conversion is minimised. The overall environmental benefits of bio-energy production on farmland will depend, however, on its impact on overall land use intensity and agricultural biodiversity, e.g. semi-natural grasslands.

To date, the most significant agricultural land use associated with renewable energy production in the EU is crop area devoted to bio diesel (mainly oilseed rape) and ethanol crops (mainly sugar beet, and cereals). In 2003, an estimated 1.6 million ha of agricultural land in the EU-15 were devoted directly to the production of renewable energy from primary biomass sources. In absolute amounts, the agriculture sector contributed 2.23 million tons of oil equivalent (M toe) of primary energy from renewable resources, which includes bio fuels (67 % of total primary energy including transportation and heating), short rotation forestry (13 %), biogas (3 %), and straw use (17 %) (Figure 7.6).

The production of bio diesel from oilseed crops has increased more than ten-fold in the period 1994–2003, resulting in a primary energy production of 1354 ktoe per year or 3.6 % of total renewable energy production in the year 2003. This development is a result of the implementation of the EU Directive on the promotion of the use of bio fuels or other renewable fuels for transport (2003/30/EC) that calls for a 5.75 % replacement of fossil transportation fuels by the year 2010. To date, however, seven EU-15 Member States have either none or negligible crop production for bio fuels, and 86 % of total bio fuel crops are produced in four EU-15 Member States.

The production of biogas from anaerobic digestion of agricultural crops and residues has increased significantly in a number of EU-15 Member States, mostly due to changes in legislation in relation to co-digestion (which makes anaerobic digestion more efficient and economically profitable) and premium payments for renewable electricity production. Direct land use related to anaerobic digestion (related to co-digestion of crops) is difficult to estimate due to lack of available data.

The Member States with a relatively large share of land devoted to energy crops include Germany,

Italy, France, and Spain. In two of these Member States (Germany, France), a significant fraction of energy crops are produced on non-food set-aside land. According to the Farm Structure Survey data, 17 % of total set-aside land in the EU-15 was used for non-food/non-feed purposes in 2000.

7.7 Conclusions: evaluation of indicators

7.7.1 Summary: general evaluation

Six of the nine indicators used in this agri-environmental storyline are classed as 'useful'. This means that these indicators are recommended to be retained for future agri-environmental indicator work. The indicators which are considered as 'potentially useful' are the response indicators ('regional levels of environmental targets' and 'production of renewable energy') as well as the driving force indicator 'energy use'.

The low scores for three indicators relate mainly to low policy relevance, slow responsiveness and weaknesses in analytical soundness. The indicator 'Regional levels of environmental targets' scores particularly low with regard to data availability and measurability (lack of time-series information and regional data).

The following sections present in more detail the evaluation of individual indicators according to the criteria set out in Section 2.3. Table 7.2 summarises the scoring for all indicators in this storyline.

7.7.2 Policy relevance

The indicators considered to be directly linked to particular Community targets, objectives or legislation are the emission indicators (IRENA 18sub, 19 and 34.1), and the indicator 'production of renewable energy' (IRENA 27). The GHG emission indicators are linked in particular to meeting the objectives of the international UN Framework Convention on Climate Change (UNFCCC). The ammonia indicator is linked to the international UNECE Protocol to Abate Acidification, Eutrophication and Ground-level Ozone (the Gothenburg Protocol), and EU directives. The production of renewable energy is linked to achieving the objectives of the Kyoto Protocol and the EU Directive on promotion of (transportation) bio fuels. These indicators are also assessed as very useful for decision makers as they indicate how close the EU-15 Member States are in reaching agreed targets.

The driving force indicators, 'energy use' (IRENA 11), 'mineral fertiliser consumption' (IRENA 8), 'cropping/livestock patterns' (IRENA 13) and 'farm practices-manure management' (IRENA 14.3) provide input to the emissions indicators, and therefore have an indirect link to policy targets and legislation related to air quality and climate change. These indicators are considered to be fairly useful, because they only contribute to one particular aspect of the issue. For instance, 'energy use' (IRENA 11) is indirectly linked to UNFCCC, but the indicator does not address CO_2 emissions from agricultural operations, or the production of fertilisers.

The response indicator on 'regional levels of environmental targets' (IRENA 3) is regarded to be policy relevant, providing an overview of the Community targets, objectives or legislation relevant to climate change and air quality.

7.7.3 Responsiveness

The emissions indicators (IRENA 18sub, 19 and 34.1) and the 'production of renewable energy' (IRENA 27) are regarded as being very responsive, as they integrate a number of factors and are sensitive to economic and political changes. The indicator on 'regional levels of environmental targets' (IRENA 3) is fairly sensitive as it monitors political changes. The driving force indicators (IRENA 8, 11, 13, 14) are also regarded as being fairly responsive as changes in farm practices and conversions from conventional to organic farming are not usually immediate.

7.7.4 Analytical soundness

Most indicators are based on emission coefficients linked to land and livestock numbers. The indicator on 'regional levels of environmental targets' (IRENA 3) is based on information contained in policy documents.

High quality statistics or data is used to underpin most indicators apart from 'mineral fertiliser consumption' (IRENA 8), 'production of renewable energy' (IRENA 27), which are based on non-harmonised data sources.

There is a strong quantitative link between the driving force indicators and the state indicators, with the exception of 'energy use' (IRENA 11), from which one can only ascertain a qualitative link because CO_2 emissions from indirect energy use are difficult to estimate. The response indicators also have a qualitative link because there is no quantitative feedback mechanism to the pressure or state/impact indicators.

7.7.5 Data availability and measurability

The driving force indicators are based on regional data, with the exception of 'mineral fertiliser consumption' (IRENA 8), for which national and EU-15 data is available. However, the pressure and State/impact indicators are all based on information at national level. This is because the UN conventions only require information to be reported at national level. The response indicators are also compiled at national level.

In terms of availability of time series, all driving force indicators have long-term data (from 1990 onwards), apart from information concerning storage facilities for manure, which was recently introduced into the Farm Structure Survey. The pressure and State/impact indicators are all underpinned by data collected from 1990 onwards. The response indicators, however, are all based on short-term data.

7.7.6 Ease of interpretation

All indicators used in the environmental story line provide messages that are very clear to understand.

7.7.7 Cost effectiveness

All indicators are based on existing statistics and data sets, with the exception of 'regional levels of environmental targets'. All data is considered to be easily accessible because they are regularly provided by Eurostat, DG Agriculture and Rural Development (in charge of FADN) or the EEA ETC for Air and Climate Change, apart from the production of renewable energy resources, which is based on a variety of different sources.

Table 7.2 Evaluation of indicators used to analyse agriculture's impact on air and climate change

Indicator criteria	Sub-criteria	Scoring	Driving forces				Pressures		Impact	Responses	
			Energy use	Livestock patterns	Mineral fertiliser consumption	Farm management (practices) — manure management	Atmospheric emissions of ammonia from agriculture	Emissions of methane and nitrous oxide from agriculture	Share of agriculture in greenhouse gas emissions	Regional levels of environmental targets	Production of renewable energy (by source)
		IRENA indicator no	11	13	8	14	18sub	19	34.1	3	27
Policy relevance	Is the indicator directly linked to Community policy targets, objectives or legislation?	0 = No 1 = Yes, indirectly 2 = Yes, directly	1	1	1	1	2	2	2	1	1
	Could the indicator provide information that is useful to policy action/ decision?	0 = Not at all 1 = Fairly useful 2 = Very useful	1	1	1	1	2	2	2	1	2
Respon-siveness	Is the indicator responsive to environmental, economic or political changes?	0 = Slow, delayed response 1 = Fast, immediate response	0	0	0	0	1	1	1	0	1

Climate change and air quality

Indicator criteria	Sub-criteria	Scoring	Driving forces				Pressures		Impact	Responses	
			Energy use	Livestock patterns	Mineral fertiliser consumption	Farm management (practices) — manure management	Atmospheric emissions of ammonia from agriculture	Emissions of methane and nitrous oxide from agriculture	Share of agriculture in greenhouse gas emissions	Regional levels of environmental targets	Production of renewable energy (by source)
		IRENA indicator no	11	13	8	14	18sub	19	34.1	3	27
Analytical soundness	Is the indicator based on indirect (or modelled) or direct measurements of a state/trend?	0 = Indirect 1 = Modelled 2 = Direct	1	2	2	2	1	1	2	2	2
	Is the indicator based on low/medium/high quality statistics or data?	0 = Low quality statistics/data 1 = Medium quality statistics/data 2 = High quality statistics/data	2	2	1	2	2	2	2	0	1
	What are the causal links with other indicators within the DPSIR framework?	0 = Weak or no link 1 = Qualitative link 2 = Quantitative link	1	2	2	2	2	2	2	1	1
Data availability and measurability	Good geographical coverage?	0 = Only case studies 1 = EU-15 and national 2 = EU-15, national and regional	2	2	1	2	1	1	1	1	1
	Availability of time series	0 = No 1 = Occasional data source 2 = Regular data source	2	2	2	1	2	2	2	0	1
Ease of interpretation	Are the key messages clear and easy to understand?	0 = Not at all 1 = Fairly clear 2 = Very clear	2	2	2	2	2	2	2	2	2
Cost effectiveness	Based on existing statistics and data sets?	0 = No 1 = Yes	1	1	1	1	1	1	1	1	1
	Are the statistics or data needed for compilation easily accessible?	0 = No 1 = Yes, but requires lengthy processing 2 = Yes	1	2	2	2	2	2	2	2	1
Total score			14	17	15	16	18	18	19	11	14
Classification of indicators: 0 to 7 (*) = 'Low potential' 8 to 14 (**) = 'Potentially useful' 15 to 20 (***) = 'Useful'			**	***	***	***	***	***	***	**	**
Final classification of indicators according to the following criteria: Policy relevance at least 2 points; Analytical soundness at least 4 points; Data availability at least 3 points			**	***	***	***	***	***	***	**	**

8 Biodiversity and landscape

8.1 Summary of main points

- Extensive farming systems are important for maintaining the biological and landscape diversity of farmland, including Natura 2000 sites. Such systems have long been threatened, however, by two different trends: intensification and abandonment.
- While intensification, in terms of the use of external inputs, seems to have levelled off during the 1990s, the trend towards farm specialisation continues in the EU-15. The decline in the proportion of 'mixed livestock' farms by about 25 % from 1990 to 2000 is particularly significant as these farms are often associated with high biodiversity and landscape quality.
- Risks for the marginalisation of farmland have been identified in Ireland, southern Portugal and large parts of Italy. This leads potentially to farm abandonment with an associated loss of high nature value farmland and characteristic agricultural landscapes.
- High nature value (HNV) farmland contains the most biodiversity-rich areas within agricultural landscapes. HNV farmland areas are mainly found in the Mediterranean region, upland areas in the United Kingdom and Ireland, mountain areas and in some parts of Scandinavia.
- The majority of farmland birds have suffered a strong decline from 1980 to 2002. This decline has levelled off in the 1990s but species diversity remains at a very low level in intensively farmed areas. Data for important bird areas and Prime Butterfly Areas show that a significant share of these sites is negatively affected by agricultural intensification and/or abandonment.
- Various policy measures can be used to tackle biodiversity decline on farmland. These include site protection, agri-environment measures, codes of good farming practice, and conversion to organic farming.
- According to current estimates about 18 % of the habitats in Natura 2000 areas depend on a continuation of extensive agricultural practices. The appropriate management of such areas by farmers can be supported through agricultural policy instruments, such as agri-environment schemes.
- The budgetary expenditure and level of area coverage by agri-environment schemes varies considerably between the EU-15 Member States. It is below average in most southern Member States, where the share of HNV farmland and Natura 2000 sites is relatively high.
- Organic farming and codes of good farming practice establish a framework for agricultural management that benefits habitat diversity and common farmland species. Sensitive or rare farmland species require additional measures for their survival, such as targeted agri-environment schemes.

8.2 Introduction

About half of the land in the EU-15 is managed by farmers, which gives agriculture an important role in the maintenance of biodiversity. Varying farming traditions, combined with specific soil and climate conditions, have resulted in diverse and highly characteristic agricultural landscapes often with a rich flora and fauna. Nevertheless, the biodiversity of Europe's farmland has declined strongly in the last few decades.

The most favourable conditions for many farmland species are created under extensive and/or traditional agricultural management. These species are vulnerable, therefore, to two opposing trends — the intensification of agriculture as well as the abandonment of agricultural land use (Bignal and McCracken, 1996). Agricultural land use in the more productive lowland areas of the EU-15 has intensified considerably during the last decades (see also Chapter 3). This has had negative consequences for biodiversity in these areas and in surrounding habitats due to the simplification of crop rotations, the use of fertilisers and pesticides to favour crops and productive grasses, the elimination of landscape elements, agricultural drainage and other factors (e.g. Potts, 1986; Pain & Pienkowski, 1996; Stoate *et al.*, 2001).

Agricultural areas with high biodiversity ('high nature value (HNV) farmland') still exist, mostly

Biodiversity and landscape

Figure 8.1 Environmental assessment of agriculture's impact on biodiversity and landscape based on the DPSIR framework

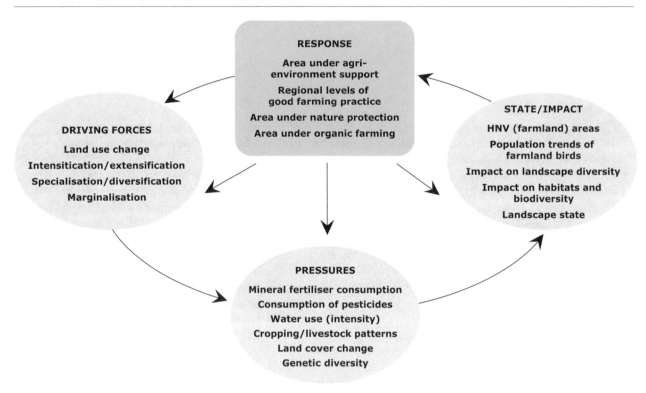

Table 8.1 IRENA indicators relevant for assessing agriculture's impact on biodiversity and landscapes

DPSIR	IRENA indicators	
Driving forces	No 12	Land use change
	No 15	Intensification/extensification
	No 16	Specialisation/diversification
	No 17	Marginalisation
Pressures	No 8	Mineral fertiliser consumption *
	No 9	Consumption of pesticides *
	No 10	Water use (intensity) *
	No 13	Cropping/livestock patterns *
	No 24	Land cover change
	No 25	Genetic diversity
State	No 26	High nature value (farmland) areas
	No 28	Population trends of farmland birds
	No 32	Landscape state
Impact	No 33	Impact on habitats and biodiversity
	No 35	Impact on landscape diversity
Responses	No 1	Area under agri-environment support
	No 2	Regional levels of good farming practices
	No 4	Area under nature protection*
	No 7	Area under organic farming

* In the context of biodiversity it appeared advisable to include the input use indicators 8, 9 and 10 in the pressures category. As cropping and livestock patterns are expressions of other driving force indicators and often impact directly on farmland biodiversity, indicator 13 is classified here as pressure indicator. Indicator 4 is used in two different contexts: to discuss the link between agriculture and biodiversity and as part of the policy response.

in mountain areas and on other less productive land. However, farmers in these areas are generally under economic pressure and may choose either to intensify or give up farming entirely. Land abandonment has occurred throughout Europe since at least the middle of the last century. Since it affects mainly extensive traditional farming systems, its consequences for biodiversity are generally regarded as undesirable (Baldock *et al.* 1996, Brouwer *et al.* 2001).

This chapter attempts to trace the positive and negative links between agriculture and biodiversity, underlying driving forces and policy responses with the help of relevant IRENA indicators.

8.3 IRENA indicators related to biodiversity and landscape

The Driving force — Pressure — State/impact — Response framework provides a means to show linkages and associations between indicators (Figure 8.1 and Table 8.1), and to structure the environmental assessment of agriculture's impact on biodiversity and landscape.

The use of inputs (IRENA 8, 9, 10) can lead to direct or indirect pressures on biodiversity. These indicators are included in the indicator set for this storyline but not evaluated in detail as direct links are difficult to establish. However, increases or decreases in the use of mineral fertilisers, pesticides and water underpin, or are closely associated with, changes in cropping and livestock patterns or intensification/extensification trends that have negative or positive impacts on farmland biodiversity. There is also strong evidence for the negative effects of pesticide use on farmland birds (Campbell and Cooke, 1997). The biodiversity impact of increases in irrigation area is briefly described in Section 4.6.2.

8.4 Trends derived from driving force indicators

Agriculture has changed considerably in the last century and is continuing to evolve. Key driving forces behind these changes are represented by the indicators on 'fertiliser consumption' (IRENA 8), 'consumption of pesticides' (IRENA 9), 'water use (intensity)' (IRENA 10), 'land use change' (IRENA 12), 'intensification/extensification' (IRENA 15), 'specialisation/diversification' (IRENA 16) and 'marginalisation' (IRENA 17).

Chapter 3 on general trends in agriculture provides the following key messages linked to biodiversity and landscapes:

- Input use increased for pesticides and water but decreased for fertilisers during the 1990s.
- For the EU the share of the agricultural area managed by high-input farms decreased from 44 % in 1990 to 37 % in 2000. Low-input farms increased their share from 26 % to 28 % of agricultural area. Both these trends are for EU-12.
- From 1990 to 2000, specialised farms increased by 4 % (from 68.7 million ha to 71.2 million ha), whereas non-specialised farms decreased by 18 % (from 33.7 million ha to 27.7 million ha).
- The proportion of 'mixed livestock' farms declined from 16 % in 1990 to 12 % in 2000. This trend has serious implications since farms (often a combination of cattle and sheep) are often associated with high biodiversity and landscape quality.
- Marginalisation of farmland appears to be occurring in Ireland, southern Portugal, Northern Ireland and large parts of Italy, leading to a risk of farm abandonment.
- During 1990 to 2000, the change in land use from agriculture to artificial surfaces ranges from 2.9 % in the Netherlands to 0.3 % in France. In general the highest percentage of agricultural land (in 1990) converted to artificial areas (in 2000) is found close to major cities.

8.5 Agricultural pressures and benefits on biodiversity and landscapes

The farm trends set out in the above section are linked to changing cropping/livestock patterns and land cover (IRENA 13 and IRENA 24) as well as to trends in the genetic diversity of crops and livestock (IRENA 25). These can have negative and positive effects on landscapes and biodiversity.

The type and mixture of agricultural land use determine habitat availability and quality for farmland species. IRENA 13 shows that arable land is the dominant agricultural land use in large parts of Europe (see Figure 3.1). Permanent grassland dominates agricultural areas in parts of lowland western Europe, mountainous areas and Scandinavian Member States. Landscapes dominated by olive groves, vineyards and other permanent crops, are mostly restricted to areas in southern Europe.

Due to urbanisation and the conversion to forest and non-agricultural land cover, the utilised agricultural area (UAA) for the EU-12 decreased by 2.5 % (2 883 520 ha) between 1990 and 2000. Permanent grassland and permanent crops decreased by 4.8 % (2 079 700 ha) and 3.8 % (389 170 ha), respectively. However, these general trends mask even stronger regional changes that have potential negative implications for biological and landscape diversity. The decline in permanent grassland area reduces the land use mosaic in

Biodiversity and landscape

Figure 8.2 National share of Natura 2000 habitats that depend on a continuation of extensive farming practices within Natura 2000 sites

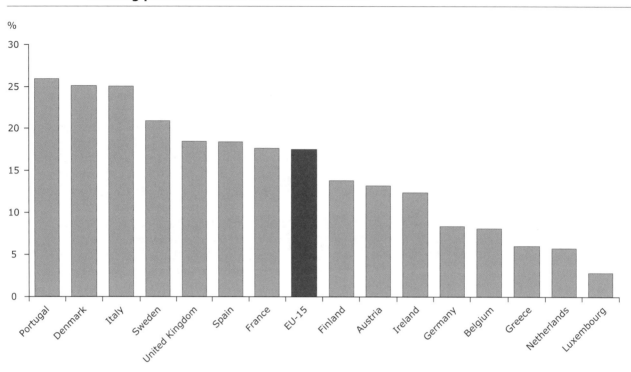

Source: Reporting of Member States in the framework of the Habitats Directive (92/42/EEC). Snapshot, July 2004.

landscapes, which are already used intensively. The decrease of permanent crops, in particular vineyards, in Mediterranean regions can reduce landscape and habitat diversity in these areas.

Cattle remains the dominant livestock type in the majority of regions, even though total cattle numbers decreased by 8.3 % between 1990 and 2000 (see Section 3.3.2). This decline was particularly marked in the midlands and southeast of the United Kingdom, southern Germany, parts of Italy and the north-west, south-west and east of France and northwest of Spain. Sheep numbers decreased by 3.4 % but sheep remain dominant in some regions of Spain, southern Greece, and upland areas and in some parts of the United Kingdom. High sheep stocking densities have led to overgrazing of fragile heather moor land in the United Kingdom uplands and need to be reduced to improve the conservation status of many protected areas (English Nature, 2003).

The combination of different livestock types in lowland areas is often the most advantageous from the perspective of biodiversity or landscape — and in particular, the combination of cattle and sheep. The decline of mixed livestock farms, as indicated by the FADN survey, is therefore of concern from a biodiversity perspective.

Data on genetic diversity (IRENA 25) for most EU-15 Member States are limited and difficult to interpret. Traditional livestock breeds are often associated with extensive grazing practices and high nature value farmland. Moreover, modern high-yielding dairy cattle especially require high-energy fodder and are therefore not suitable for grazing semi-natural grasslands. Thus, there is a need to assess trends in the genetic diversity of crops and livestock. FAO data show that about 50 % of the main livestock breeds (cattle, pig, sheep, goat and poultry) in EU-15 Member States are either extinct or have an endangered or critical status. The information available reflects partly different approaches to reporting livestock breed data (in terms of time series, type of breed covered etc.) so that comparisons between countries or time-trend analyses are not possible.

The decline of traditional livestock breeds has negative implications for the management of semi-natural habitats that have been shaped by traditional agricultural practices, including many future Natura 2000 areas. Figure 8.2 shows that, on average, 18 % of all land in Natura 2000 areas belongs to habitat categories, which depend on a continuation of extensive farming practices (IRENA 4).

Figure 8.3 Share of high nature value (HNV) farmland areas in total UAA

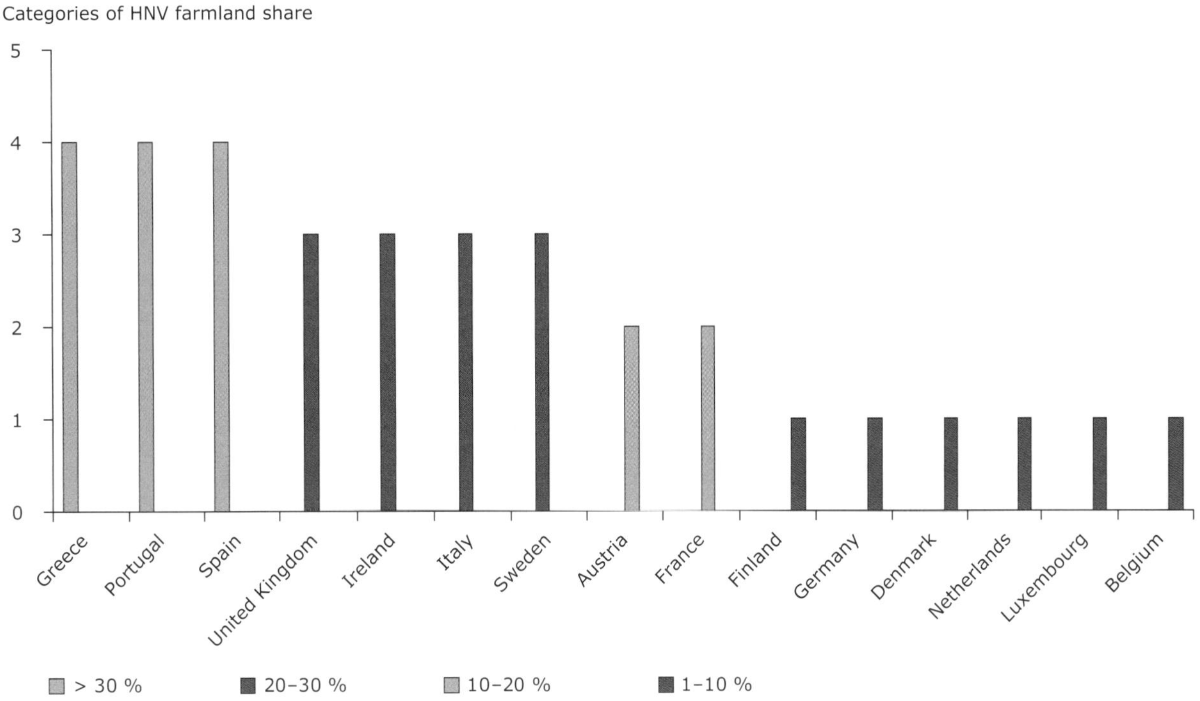

Categories of HNV farmland share

Greece: 4, Portugal: 4, Spain: 4, United Kingdom: 3, Ireland: 3, Italy: 3, Sweden: 3, Austria: 2, France: 2, Finland: 1, Germany: 1, Denmark: 1, Netherlands: 1, Luxembourg: 1, Belgium: 1

☐ > 30 %   ■ 20–30 %   ☐ 10–20 %   ■ 1–10 %

Source: EEA, 2004c.

8.6 State of/impacts on biodiversity and landscape

8.6.1 Biodiversity

The previous section has shown that extensive agricultural management is important for the creation and maintenance of farmland habitats in the EU-15. However, changes in modern farming systems and farm management lead to significant pressures on biological and landscape diversity. Even though data for the EU-15 Member States are not always fully comparable or complete, several indicators provide information on the state of, and impact of agriculture on, biodiversity and landscapes.

High nature value (HNV) farmland areas (IRENA 26), which contain the most biodiversity-rich areas of farmland, are mainly found in Mediterranean regions, upland areas in the United Kingdom and Ireland, mountain areas and parts of Scandinavia (EEA, 2004c). The share of high nature value farmland in each EU-15 Member State can be estimated by combining data derived from the share of HNV farm types and land cover classes that are associated with high agricultural biodiversity. Both approaches have their specific weaknesses but the estimated share of HNV farmland area at EU-15 level is 15–25 % of the total utilised agricultural area. Current data do not allow assessing trends in the share of HNV farmland for individual Member States but give an overall impression of the share of such areas in EU-15 Member States (Figure 8.3).

There are few data about the actual conservation status or species diversity of high nature value farmland. However, information collected by voluntary organisations on important bird areas (IRENA 28) and the distribution of rare and threatened butterflies (see IRENA 33) gives an indication of biodiversity trends on HNV farmland.

Figure 8.4 shows the distribution of important bird areas (IBAs) that are considered as threatened by agricultural intensification according to the information published by BirdLife International (Heath and Evans, 2000). The figure does not distinguish sites on the basis of the degree of negative impact or proportion of the site affected but provides a good overview of their geographical distribution. The overlap with potential HNV farmland areas is strong in the south and west. However, many IBAs in areas of centre-west EU-15 Member States that are currently not identified as HNV farmland are also under pressure from intensification.

Rarer and more vulnerable farmland species are still in decline, as shown by the number and conservation status of butterfly species in the prime

Biodiversity and landscape

Figure 8.4 Important bird areas classified as threatened by agricultural intensification

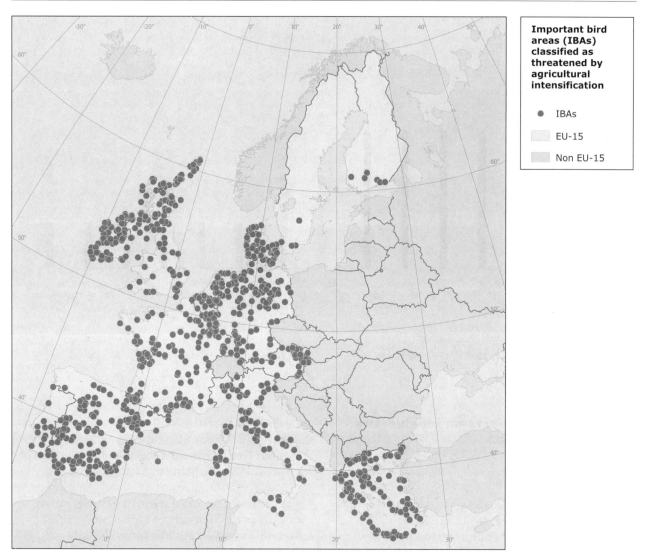

Source: BirdLife International, 2004.

butterfly areas (Figure 8.5). Their conservation status is generally negative in the EU-15, with Spain and Greece as positive exceptions. Given the close association of these butterfly species with specific farmland habitats (in particular extensive grassland), changes in agricultural land use are considered to be the main factors for their decline or positive conservation status (Van Swaay & Warren, 2003). According to Butterfly International about 40 % of all agricultural prime butterfly areas experience negative impacts from intensification and abandonment, respectively. This again shows the importance of a continuation of extensive agricultural practices for the survival of sensitive farmland species.

Many case studies document the impact of agricultural practices on flora and fauna, but the best data in terms of time series and geographic distribution are available for birds. Some common species of farmland birds have shown a dramatic decline parallel to an increase of agricultural intensity (Donald et al., 2001, Vickery et al., 2001; Holes et al., 2005). Figure 8.6 shows farmland bird population trends from 1990 when this decline levelled off. This may partly be explained by the observed stabilising of farming intensity, but in some intensive areas farmland bird diversity may have reached such low levels that it is not really affected by further intensification (BirdLife International, 2004). The population decline not only affects rare species but also (previously) common farmland birds, such as the skylark or the grey partridge (cf. Potts, 1986).

8.6.2 Landscapes

Europe has a great variety of agricultural landscapes that reflect differences in biophysical

Figure 8.5 Population trends of agriculture-related butterfly species in prime butterfly areas

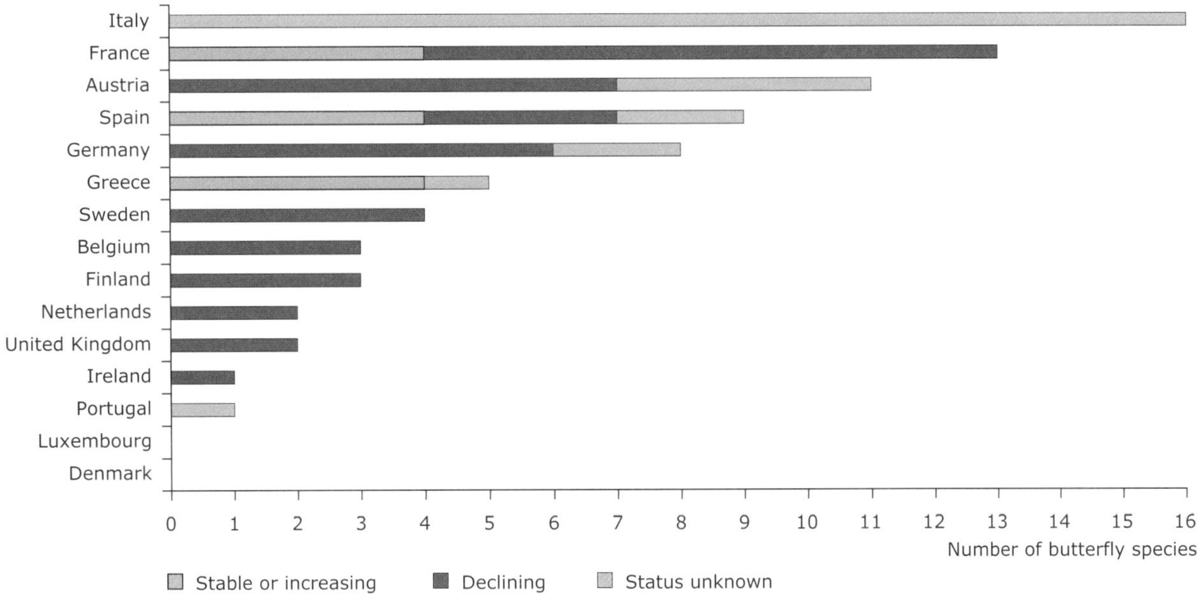

Country	Stable or increasing	Declining	Status unknown	Total
Italy			16	16
France	4	9		13
Austria	7		4	11
Spain	4	3	2	9
Germany		6	2	8
Greece		4	1	5
Sweden		4		4
Belgium		3		3
Finland		3		3
Netherlands		2		2
United Kingdom		2		2
Ireland		1		1
Portugal			1	1
Luxembourg				0
Denmark				0

Note: Denmark and Luxembourg do not retain key butterfly species that rely on agricultural habitats.

Source: Prime butterfly areas in Europe, Van Swaay and Warren, 2003.

conditions, farm management practices and cultural heritage. Farmers play a crucial role in shaping and maintaining these landscapes. Given data limitations, selected case study areas (IRENA 32),

Figure 8.6 Trend in farmland bird population index from 1990–2002 in EU-11 ([44])

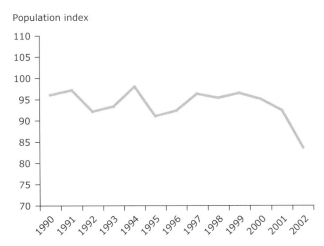

Source: BirdLife International, 2004 (Pan-European Bird Monitoring Project (RSPB/EBCC/BirdLife International/Statistics Netherlands).

representative of different landscapes, are used to show the importance of agriculture in terms of land cover in selected landscape types. Agricultural land is most dominant in the bocage (hedgerow) landscapes (84 % of total land area) and least dominant in the alpine (24 % of total land area) case study areas (Figure 8.7). However, agriculture remains important, even in alpine regions, for characterising the landscape by opening up the original forest cover.

There is great variation between different landscapes in terms of the distribution of arable, grassland, permanent crops and other agricultural land uses. Around 60 % of the land surface is covered by arable land in the open field areas landscapes of Castilla y León and eastern Denmark. Grasslands cover half of the territory in the dehesas of Extremadura, the bocage landscape in Normandy and uplands in Ireland. Permanent crops represent one-third of the agricultural land in the Montados landscape (southern Portugal), while these are non-existent in the uplands in the United Kingdom and Ireland.

The approach chosen for developing the landscape indicator is linked deliberately to capturing the importance of farming to many landscapes in the

---

[44] Bird population index trend data is obtained from the EU-15 Member States except Finland, Greece, Luxembourg, and Portugal.

Biodiversity and landscape

Figure 8.7 Percentage of agricultural land cover type in total land area for selected case study areas

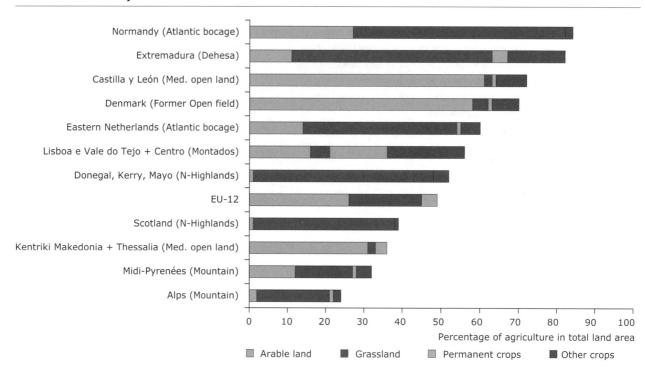

Source: Community Survey on the Structure of Agricultural Holdings (FSS), Eurostat.

EU-15. Currently, there is insufficient information and analytical possibilities for mapping European landscapes in their full complexity and range of functions.

IRENA 35 assesses the impact of changes of agricultural land use on landscapes. Between 1990 and 2000 the largest increase in grasslands (10 %) occurred in the Mediterranean open field region of Castilla y León. Conversely, in the Atlantic bocage region of Normandy, grassland decreased by 10 %, while arable land increased by 4 % during the last decade. The area of permanent crops decreased by 5 % in the Montado case study region of Portugal.

These land use changes are difficult to translate directly into environmental impacts. However, national data from the Swedish and UK Countryside surveys show that the development of linear features depends on the area surveyed. In intensively used landscapes the density of linear features seems to have declined further from 1990 to 1998. However, in more extensively farmed areas of the United Kingdom their density has sometimes increased considerably. This may be due to agri-environment schemes and other environmental measures.

8.7 Responses

A range of policy measures is available to address the decline of biological and landscape diversity on farmland, which have been translated into certain 'response indicators' related to the 'public policy' dimension. These include site protection (Natura 2000 — IRENA 4), incentives for environmental management (agri-environment schemes — IRENA 1), the setting of minimum environmental standards for farm management (good farming practice — IRENA 2), and the promotion of farming systems that support biodiversity (e.g. organic farming — IRENA 7). These four different policy responses are briefly evaluated in the following section.

8.7.1 Natura 2000 sites and extensive farming practice

Section 8.6 has explained the importance of extensive farming practices for the maintenance of biological diversity on farmland. This relationship also holds in protected areas that have been, or are proposed to be, designated under the birds and habitats directives. Figure 8.8 shows the proportion of candidate Natura 2000 sites (pSCIs) covered by targeted habitats

Biodiversity and landscape

Figure 8.8 Regional share of Natura 2000 habitats that depend on a continuation of extensive farming practices within Natura 2000 sites

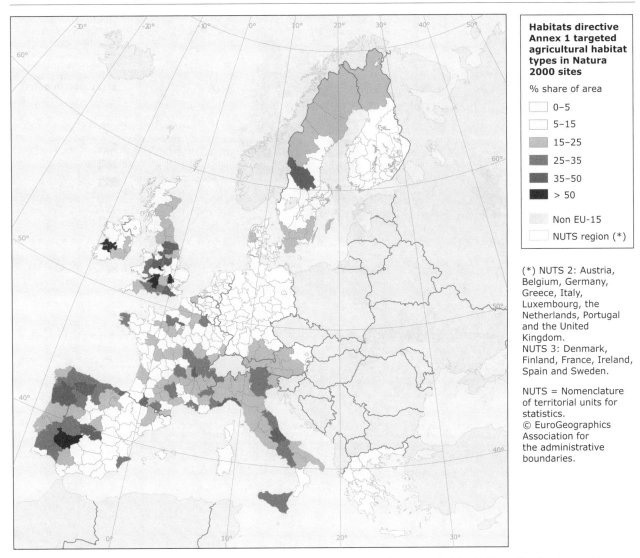

Note: The analysis focuses on agricultural Natura 2000 habitats that are listed in Annex I of the Habitats Directive (92/43/EC).
Source: Natura 2000, DG Environment, 2004. Snapshot, July 2004.

that require extensive farming practices for a favourable conservation status (IRENA 4, Area under nature protection). The legal requirements associated with site protection are an important tool to prevent damage to biodiversity from agricultural activities. Where the continuation of positive farming practices is uneconomic to farmers, however, the active management of important habitats needs to be supported by additional measures. Agri-environment schemes and other CAP policy instruments are likely also to play a key role, therefore, in maintaining the conservation status of the future Natura 2000 network.

8.7.2 Area under agri-environment support

Agri-environment measures are specifically aimed at achieving positive environmental management. EU Member States can grant support to farmers for a range of environmentally favourable measures, including biodiversity related measures and conservation of high nature value farmland. The total area of agri-environment schemes in 2002 amounted to nearly 30.2 million ha in the EU-15. The share of agricultural land enrolled in agri-environment measures in total UAA has increased from approximately 20 % in 1998 to 24 % in 2002. In Finland, Sweden, Luxemburg and Austria large proportions of the utilised agricultural area are under agri-environment schemes (more than 75 %), in contrast with Belgium, the Netherlands, Spain and

Biodiversity and landscape

Figure 8.9 Share of utilised agricultural area under agri-environment schemes in 2001

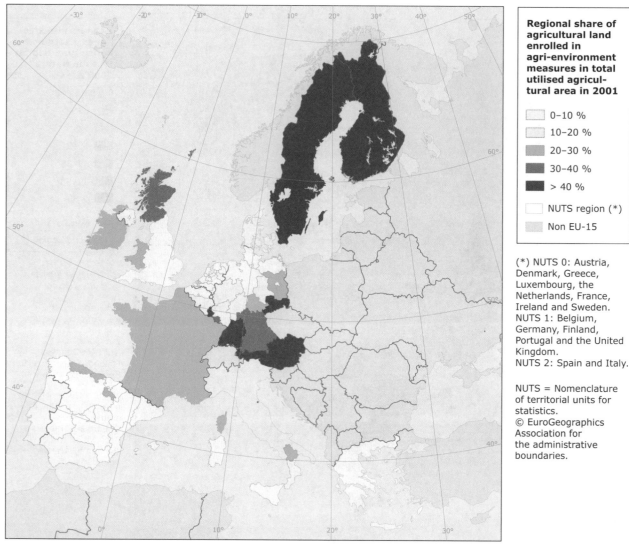

Source: DG Agriculture and Rural Development. Common indicators for monitoring the implementation of rural development programmes 2001. Includes agri-environmental contracts under the predecessor Regulation (EC) 2078/1992 and contracts signed in 2000 and 2001 under the current Regulation (EC) 1257/1999.

Greece where the area enrolled amounts to less than 10 % of UAA (Figures 8.9 and 8.10).

The relatively low coverage in southern European Member States in particular means that the potential of this policy instrument for the conservation of HNV farmland and probably also Natura 2000 sites may not be fully utilised (EEA, 2004c). The potential benefit of agri-environment schemes for biodiversity is also limited by the share of agri-environment contract area targeted on landscape and biodiversity. The schemes specifically targeted at nature and landscape enhancement represent 30 % (8.1 million ha). This type of commitment includes all actions that aim at the conservation, restoration and creation of nature (e.g. biotopes, field margins, wetlands etc.).

Figure 8.10 shows that the coverage of utilised agricultural area by agri-environment schemes varies considerably between Member States. This also applies to the share and total extent of nature and landscape oriented schemes. Sweden and Austria put a strong focus this type of agri-environment scheme. The other Member States that exceed the EU-15 average are Belgium, Portugal, Spain and United Kingdom, whereas targeted nature and landscape measures cover only a very small area in the remaining Member States. It is not only total area covered, however, that counts

Biodiversity and landscape

Figure 8.10 Share of farmland (UAA) under agri-environment schemes in 2002

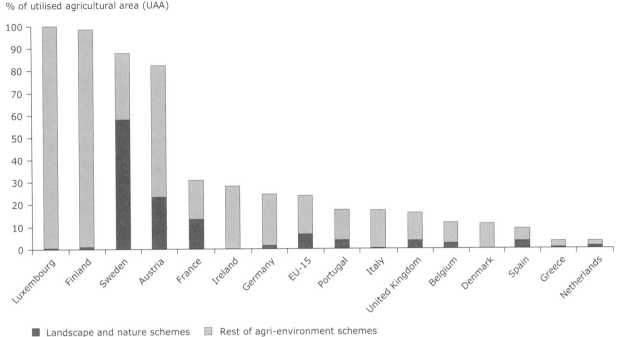

■ Landscape and nature schemes   □ Rest of agri-environment schemes

Note: The 'landscape and nature' category only includes the area under new agri-environmental contracts signed in 2000–2002 under Regulation (EC) 1257/1999 (equivalent to a total of 8.2 million ha) because the data is not available for the 'old' schemes.

Source: DG Agriculture and rural development, Common indicators for monitoring the implementation of rural development programmes, 2002. For Italy, data come from INEA.

but also the suitability of schemes. Experiences in the United Kingdom show that targeted agri-environment schemes can reverse the decline of bird species (e.g. Vickery et al., 2004). It should be noted that individual Member States, such as the Netherlands, finance nature-oriented agri-environment measures from their national budget.

8.7.3 Regional levels of good farming practice

Codes of good farming practice (GFP) are another policy tool that will encourage, among other objectives, the preservation of biodiversity and agricultural landscapes. Member States have to define codes of good farming practice at national or regional level in their rural development programmes (RDPs). Codes of good farming practice constitute the baseline requirements for farmers wishing to join agri-environment schemes. Farmers receiving compensatory allowances in Less Favoured Areas and areas with environmental restrictions (mainly sites in relation with habitats and birds directives) are also required to respect the standards of GFP.

Table 8.2 Degree of coverage of relevant categories of farming practices by national codes of GFP

Farming practices	BE-Fl	BE-Wa	DK	DE	GR	ES	FR	IE	IT-ER	LU	NL	AT	PT	FI	SE	UK
Biodiversity and Landscape	□	□	□	—	■	□	□	■	—	□	—	□	□	□	□	■
Pasture management	—	□	—	—	□	□	■	■	—	■	—	—	□	—	—	■

■ Priority issue   — Issue not covered   □ Issue addressed

Source: Compiled from codes of GFP described in national rural development programmes 2000–2006.

Biodiversity and landscape

Figure 8.11 Share of agricultural land under organic farming in 2000

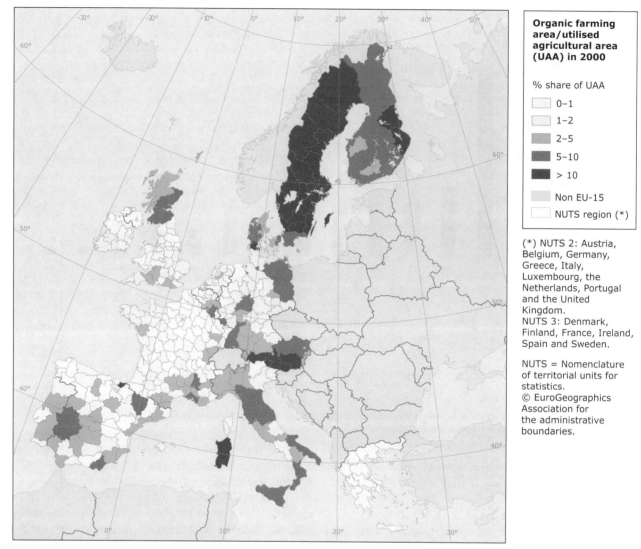

Source: Community Survey on the Structure of Agricultural Holdings (FSS), Eurostat.

According to IRENA 2, the majority of Member States, apart from Germany, Italy (the analysis is based on the code of the region Emilia Romagna) and the Netherlands include provisions of relevance for biodiversity and landscape protection in their codes of GFP (Table 8.2).

Limit requirements on stocking density to avoid overgrazing and undergrazing are set out in Spain, Portugal, Greece and France. Some recommendations for maintaining uncultivated strips in field boundaries and hedgerows are provided in Portugal, Greece and Luxembourg. In France and Sweden the codes only stipulate not to destroy biotopes (as specified in the legislation) without referring directly to the restriction to which farmers are subjected. However, in Sweden these requirements apply only as eligibility criteria for some specific agri-environment schemes related to the conservation of biodiversity and cultural heritage values.

8.7.4 Area under organic farming

Support for organic farming through payments under agri-environment schemes is a key response at EU and Member State level for promoting farming approaches that minimise the impact of agriculture on the environment. Also, the Commission has recently adopted a European action plan for organic food and farming which is matched by national action plans in many Member States (see IRENA 3).

Research papers show that organic farming provides benefits to landscape and biodiversity, for example through a higher diversity of wildlife habitats

(Stolze et al., 2000). This conclusion is supported by Hole et al. (2005), who carried out a comparative assessment of 76 literature studies. They conclude that organic agriculture performs better than conventional agriculture both with regard to botanic species-richness and bird abundance. The study shows, however, that for some invertebrates such as earthworms, butterflies, spiders and beetles the trend is not always as clear.

Since organic farming can still be very productive and intensive (albeit not in terms of chemical inputs), its benefits for biodiversity conservation apply mainly to more generalist species. The more critical farmland habitats often require a specific management that goes beyond the standards of organic farming. Nevertheless, it is a very useful contribution to raising general environmental conditions from which many farmland species can benefit. The conversion to organic farming can also provide considerable economic advantages to low-input farming systems in marginal areas of Europe that are associated with high nature value farmland (see IRENA 7).

Organic farming is particularly widespread in Sweden and Austria and also relatively common in Finland, Denmark, Italy as well as some regions of Spain and Scotland (Figure 8.11).

8.8 Conclusions: evaluation of indicators

8.8.1 Summary: general evaluation

Half (eight out of sixteen) of the indicators in this agri-environmental storyline are classed in the category 'useful'. These are the driving force indicators: 'land use change' (IRENA 12), 'intensification/extensification' (IRENA 15), 'specialisation/diversification' (IRENA 16), the pressure indicators 'cropping/livestock patterns' (IRENA 13) and 'land cover change' (IRENA 24), the state indicator 'population trends of farmland birds' (IRENA 28), and the response indicators 'area under nature protection' (IRENA 4) and 'area under organic farming' (IRENA 7).

The indicators 'marginalisation' (IRENA 17), 'genetic diversity' (IRENA 25), 'landscape state' (IRENA 32), 'impact on landscape diversity' (IRENA 35), 'area under agri-environment support' (IRENA 1), and 'regional levels of good farming practice' (IRENA 2) are considered as 'potentially useful'.

None of the indicators were classified as low potential which means that all indicators can be recommended to be retained in future agri-environment work. The domains that are clearly weaker than the others are the state and impact indicators, which score lower on the availability of regional and time series data.

The following sections present in more detail the evaluation of individual indicators according to the criteria set out in Section 2.3. Table 8.3 summarises the scoring for all indicators in this storyline.

8.8.2 Policy relevance

The indicators considered to be directly linked to particular Community targets, objectives or legislation are IRENA 1, 2, 4, 7, 25, 28 and 33. These indicators are linked in particular to the EU objective to stop biodiversity loss by 2010 (IRENA 28, 33), but also to rural development policy, which offers measures to support the conservation and enhancement of agricultural biodiversity (IRENA 1 and 2). The landscape indicators (IRENA 32 and 35) can be linked to the Pan-European Biological and Landscape Diversity Strategy, which is a European response to support implementation of the Convention on Biological Diversity. For this reason the landscape indicators are regarded as being 'fairly useful' to policy makers. 'Land use change' (IRENA 12) is indirectly linked to initiatives on spatial planning, for example, the European Spatial Development Perspective (ESDP) (European Commission, 1999) and the Commission Communication 'Towards a thematic strategy on the urban environment' (COM (2004) 60). In addition, the Commission's proposal for a Thematic Strategy for Soil Protection (COM (2002) 179) recognises that soil sealing is a threat to conserving the beneficial effects of soil functions. 'Land cover change' (IRENA 24) is indirectly linked to a number of policy initiatives, including the habitats and birds directives, the water framework directive, the European Spatial Development Perspective, and the European Spatial Observatory Network, as well international agreements such as the Ramsar Convention, and the Convention on Biological Diversity.

The other driving force and pressure indicators are also fairly useful because they provide information on the factors that may influence biodiversity and landscape.

8.8.3 Responsiveness

The indicators on 'cropping/livestock patterns', and 'area under agri-environment support' are the indicators that are considered to be most sensitive to economic, political and environmental changes.

Biodiversity and landscape

IRENA 2 ('regional levels of good farming practice') responds only slowly to external changes as it reflects medium to long term policy processes. The indicators related to the state of biodiversity, such as 'genetic diversity', 'population trends of farmland birds' (IRENA 28) and 'high nature value farmland areas' (IRENA 26) are less responsive, as they involve biological processes. In addition, the 'intensification/extensification', 'specialisation/diversification', 'marginalisation' and the Corine land cover based indicators have a slow or delayed response.

8.8.4 Analytical soundness

All indicators are based on direct measurements, apart from the indicators 'marginalisation' (IRENA 17) and 'high nature value farmland areas' (IRENA 26), which combine different data sets and expert knowledge. Further conceptual development is required for the landscape-related indicators (IRENA 32 and 35) that tackle issues of great complexity.

The indicator on 'genetic diversity' (IRENA 25) is the only one underpinned by low quality data, relatively scattered and not harmonised. High quality statistics or data are used to compile most of the driving force, pressure and State/impact indicators, while the indicators IRENA 1, 2, 4, 25, 26, 28 and 33 are considered as relying of medium quality statistics.

All indicators are considered to have a strong qualitative link to others within the DPSIR framework.

8.8.5 Data availability and measurability

The driving force and pressure indicators are all based on regional and long term time series data, with the exception of 'genetic diversity' (IRENA 25), which builds only on some national (i.e. not all EU-15 Member States) and short term time-series data. However, the State/impact indicators are either based on national information or case study data only, and no regular data sources allowing time series exist for most of them. The landscape indicators (IRENA 32 and 35) are underpinned by regular data sources (e.g. FSS and CLC) but are based on case studies due to the difficulty of delineating different landscape types across Europe. The indicator 'area under agri-environment support' would be more useful if data was available for NUTS 2/3 regions in all Member States.

8.8.6 Ease of interpretation

All indicators used in the story line provide messages that are easy to understand, apart from 'marginalisation', 'genetic diversity', the two landscape indicators and the indicator on habitats and biodiversity. In the cases of the last two, this is because the underlying data is insufficient to give a clear message. In the case of marginalisation, the indicator combines social and economic data to derive the share of farms, which are at risk of marginalisation, which has only a qualitative link to possible farm abandonment. The other indicators are not underpinned by sufficiently good data to derive a clear message.

8.8.7 Cost effectiveness

All indicators are based on existing statistics or data sets. However, some data is not part of public data sets because they are collected by private organisations, such as the trends in farmland bird and butterfly populations. The indicators on land use change (IRENA 12), land cover change (IRENA 24) and area under nature protection (IRENA 4) need substantial processing of underlying data sources, which means that these indicators are not easy to compile in a systematic way.

Table 8.3 Evaluation of indicators used to analyse agriculture's impact on biodiversity and landscape

Indicator criteria	Sub-criteria	Scoring	Driving forces				Pressures			State/impact					Responses			
			Land use change	Intensification/extensification	Specialisation/Diversification	Marginalisation	Cropping/livestock patterns	Land cover change	Genetic diversity	Population of farmland birds	High nature value (farmland) areas	Landscape state	Impact on habitats and biodiversity	Impact on landscape diversity	Area under agri-environment support	Regional levels of good farming practices	Area under nature protection	Area under organic farming
	IRENA indicator no		12	15	16	17	13	24	25	28	26	32	33	35	1	2	4	7
Policy relevance	Is the indicator directly linked to Community policy targets, objectives or legislation?	0 = No 1 = Yes, indirectly 2 = Yes, directly	1	1	1	1	1	1	2	2	1	1	2	1	2	2	2	2
	Could the indicator provide information that is potentially useful to policy action/decision?	0 = Not at all 1 = Fairly useful 2 = Very useful	1	1	1	1	1	1	1	2	2	1	2	1	2	1	2	2
Responsiveness	Is the indicator responsive to environmental, economic or political changes?	0 = Slow, delayed response 1 = Fast, immediate response	0	0	0	0	1	0	0	1	0	0	1	0	1	0	1	1
Analytical soundness	Is the indicator based on indirect (or modelled) or direct measurements of a state/trend?	0 = Indirect 1 = Modelled 2 = Direct	2	2	2	1	2	2	2	2	1	2	2	2	2	2	2	2
	Is the indicator based on low/medium/high quality statistics or data?	0 = Low quality statistics/data 1 = Medium quality statistics/data 2 = High quality statistics/data	2	2	2	2	2	2	1	1	1	2	1	2	1	1	1	2
	What are the causal links with other indicators within the DPSIR framework?	0 = Weak or no link 1 = Qualitative link 2 = Quantitative link	1	1	1	1	1	1	1	1	1	1	1	1	1	1	1	1

Biodiversity and landscape

Indicator criteria	Sub-criteria	Scoring	Driving forces				Pressures			State/impact					Responses			
			Land use change	Intensification/extensification	Specialisation/Diversification	Marginalisation	Cropping/livestock patterns	Land cover change	Genetic diversity	Population of farmland birds	High nature value (farmland) areas	Landscape state	Impact on habitats and biodiversity	Impact on landscape diversity	Area under agri-environment support	Regional levels of good farming practices	Area under nature protection	Area under organic farming
	IRENA indicator no		12	15	16	17	13	24	25	28	26	32	33	35	1	2	4	7
Data availability and measurability	Good geographical coverage?	0 = Only case studies 1 = EU-15 and national 2 = EU-15, national and regional	2	2	2	2	2	2	1	1	1	0	1	0	1	1	2	2
	Availability of time series	0 = No 1 = Occasional data source 2 = Regular data source	2	2	2	2	2	2	1	2	1	2	0	2	1	0	1	1
Ease of interpretation	Are the key messages clear and easy to understand?	0 = Not at all 1 = Fairly clear 2 = Very clear	2	2	2	1	2	2	1	2	2	1	2	1	2	2	2	2
Cost effectiveness	Based on existing statistics and data sets?	0 = No 1 = Yes	1	1	1	1	1	1	1	1	1	1	1	1	1	0	1	1
	Are the statistics or data needed for compilation easily accessible?	0 = No 1 = Yes, but requires lengthy processing 2 = Yes	1	1	1	1	2	1	1	0	1	1	0	1	1	0	2	2
Total score			15	15	15	13	17	15	12	15	12	12	13	12	15	10	17	18
Classification of indicators: 0 to 7 (*) = 'Low potential' 8 to 14 (**) = 'Potentially useful' 15 to 20 (***) = 'Useful'			***	***	***	**	***	***	**	***	**	**	**	**	***	**	***	***
Final classification of Indicators according to the following criteria: Policy relevance at least 2 points; Analytical soundness at least 4 points; Data availability at least 3 points			***	***	***	**	***	***	**	***	**	**	**	**	**	**	***	***

9 Evaluating agri-environmental indicators and supporting data sets in the EU-15

9.1 Introduction

A key purpose of the IRENA indicator report is the evaluation of a set of 35 agri-environment indicators and their underlying data sets. This chapter summarises the results of the indicator evaluation exercise on the basis of the framework developed in Chapter 2. This exercise and the agri-environmental indicator analysis presented in Chapters 3–8 informs a review of the data set used in building the IRENA indicators. Finally, recommendations for future indicator-based monitoring and reporting are developed.

When evaluating indicators and their data sets, some basic issues that are important for their usefulness and future development have to be considered. Firstly, the development of indicators always takes place in a certain tension between the availability of statistical data and the need to analyse the relevant environmental issues in question as accurately as possible ('data-driven' versus 'problem-driven' approaches). In most cases, a combination of both approaches is needed, as it has been done in the two European Commission communications on agri-environmental indicators (COM (2000) 20 and COM (2001) 144). Secondly, even well designed indicators only give an insight into 'real-life' processes or causal relationships- they cannot fully represent them. Thus, indicator-based environmental analysis needs to be complemented by further background information and scientific study.

Furthermore, indicator development relies on the availability and accessibility of statistical data sets. The poor quality and insufficient coverage of these data sets are also important constraints to indicator development and their use in policy analysis. When evaluating the usefulness of any statistical source/tool it is sometimes worth going back to the basic question of 'what are the data used for?'. The use of statistical data for analytical purposes is obviously an important consideration in building statistical systems. The following list illustrates the potential purpose of data collection. Statistical data is used for:

- assessing trends in the object of study;
- discovering the spatial distribution of a phenomenon;
- comparative analysis between Member States, regions or issues;
- finding causal links between different observations;
- analysing policy effectiveness and targeting of measures;
- building models, scenarios and forecasts.

The complexity of analysis increases in this list from top to bottom. The last three tasks generally require a combination of different data sets and methodological approaches. During the establishment of most of the discussed data sets, the last two or three issues were not key concerns. Most data sets are, therefore, not designed for indicator-based environmental assessment work. Nevertheless, requirements of agri-environmental analysis need to inform any evaluation of data sets that underpin indicator development.

9.2 Developing and evaluating agri-environmental indicators

The development of agri-environmental indicators encounters difficulties in reality:

- Environmental issues are often too complex to be represented by individual parameters (e.g. landscape diversity),
- The territory of the European Union is very diverse as regards the farm structures (e.g. type of crops and livestock), soil characteristics, topographic and climatic conditions, farm size and agricultural productivity,
- The relationship between agriculture and environment is highly complex, to the extent that a simplified description is not necessarily helpful; the impact of many agricultural processes depends on a range of site-specific characteristics,
- Lack of or insufficient data sets prevent/limit the implementation of the most appropriate indicator concepts/methodologies, for example irrigable area has to be used as a proxy for water use,
- Causal links are not necessarily sufficiently understood to be able to represent them via indicators.

Despite these problems, agri-environmental indicators remain key instruments for environmental reporting in agriculture (and other fields). Limited resources for data collection and analysis make it necessary, however, to select a limited set of indicators that can be maintained over the longer term as part of an agri-environmental information system. This requires an evaluation of indicators from a methodological and policy relevance perspective. The report provides an assessment of the relevance and usefulness of the more than 35 investigated indicators and their linkages ('story lines') for monitoring the state of and trends in the environmental conditions in agriculture, as well as policy and sector responses. At the end of each chapter the indicators used in the relevant agri-environmental storyline were evaluated in terms of **actual** usefulness via a set of criteria based on COM (2001) 144 (see Chapter 2 for methodology). The following main conclusions arose from the evaluation exercise for the different indicator groupings.

- **General trends in agriculture** — five out of thirteen of the indicators used to show agricultural trends are classed in the category 'useful', while the rest is ranked as 'potentially useful'. In general, the indicators based on FSS, FADN and CLC scored the highest, because these sources provide harmonised regional information. However, it is difficult to link indicators reported at different scales; for example, national data on mineral fertiliser consumption (IRENA 8) with regional data on cropping and livestock patterns (IRENA 13) and regional data on yields (IRENA 15).
- **Agricultural water use** — six indicators are regarded as 'potentially useful' and one has 'Low potential' (the indicator 'ground water levels'). Better data on trends in ground water levels would be very useful but EU-level data are not available and national level data sets are very expensive to acquire. Pressure, State/impact and response indicators are underpinned by low or medium quality data and there are weak links between the indicators. Greater efforts are required to improve the indicators throughout the DPSIR framework to increase the possibilities of monitoring the impact of agriculture on water resources. Modelling may have a role to play whereby climatic information is combined with crop and land use data to determine water requirements from agriculture.
- **Agricultural input use and state of water quality** — the indicators classed as 'useful' are: 'mineral fertiliser consumption' (IRENA 8) and 'cropping/livestock patterns' (IRENA 13) and 'area under organic farming' (IRENA 7). The other eight indicators are classed in the category 'potentially useful', including 'gross nitrogen balances' that is not available at regional level. In most cases these indicators have not reached a level of development to be considered as 'useful', because data availability and measurability, and analytical soundness are inadequate. Information on the use and impact of pesticides is in particular difficult to obtain. None of the indicators are, however, regarded to have low potential.
- **Agricultural land use, farm management practices and soils** — four indicators in this environmental storyline are classed in the category 'useful': the driving force indicators 'land use change' (IRENA 12), and 'cropping/livestock patterns' (IRENA 13), the pressure indicators 'land cover change' (IRENA 24) and the response indicator 'area under organic farming' (IRENA 7). The rest of the indicators are classed in the category 'potentially useful'. This means that some of the indicators have not reached a level of development to be considered as useful, mainly due to weaknesses in data availability and measurability as well as analytical soundness. Several of them are obtained via modelling or indirect data and efforts are recommended to improve those models to achieve higher robustness and acceptability. To ensure comparable quality between the indicators the state indicators would have to be improved considerably. However, none of the indicators are regarded as 'Low potential'. 'Farm management practices' (tillage methods) (IRENA 14.1) has the lowest score. Information about tillage practices is highly relevant to soil conservation, but little reliable information is available.
- **Impact of agriculture on air and climate change** — most of the indicators (six of the nine) used in this agri-environmental storyline are classed in category 'useful'. The indicators with the highest score are those related to emissions, as 'atmospheric emissions of ammonia' (IRENA 18sub), 'emissions of methane and nitrous oxide' (IRENA 19), as well as the 'share of agriculture in GHG emissions' (IRENA 34.1). The response indicators (regional levels of environmental targets and production of renewable energy) are considered as 'potentially useful'. To become useful, their measurability needs to be improved. The 'regional levels of environmental targets' scores low because time series information is not included and it does not actually report regional information. The generally high scores of indicators in this storyline are probably linked to the fact that the

reporting level of pressure and state indicators is national and not regional, and because the pressure/State/impact indicators are largely target-driven.

- **Impact of agriculture on biodiversity and landscape** — Half (eight out of sixteen) of the indicators in this agri-environmental storyline are classed in the category 'useful'. These are the driving force indicators: 'land use changes' (IRENA 12), 'intensification/extensification' (IRENA 15), 'specialisation/diversification' (IRENA 16), the pressure indicator 'cropping/livestock patterns' (IRENA 13) and 'land cover change' (IRENA 24), the state indicator on 'populations of farmland birds' (IRENA 28), and the response indicators 'area under nature protection' (IRENA 4), area under agri-environmental support (IRENA 1) and 'area under organic farming' (IRENA 7).

The indicators 'marginalisation' (IRENA 17), 'genetic diversity' (IRENA 25), 'landscape state' (IRENA 32), 'impact on landscape diversity' (IRENA 35), 'area under agri-environment support' and 'regional levels of good farming practice' (IRENA 2), are considered as 'potentially useful'.

None of the indicators are regarded as 'Low potential'; this means that all indicators can be recommended to be retained in future agri-environment work. The domains that are clearly weaker than the others are the state and impact indicators, which score lower on the availability of regional and time series data.

The indicator on area under nature protection (IRENA 4) has a strong qualitative link to the Driving forces, Pressures, State, Impact indicators and has the potential to provide strong quantitative links, if future reporting procedures for Natura 2000 sites are designed appropriately.

An overall review of the indicator classification shows a significant influence of the data sets underpinning different IRENA indicators on their evaluation score. Data sets in the agricultural domain provide full geographic coverage, time series information and generally high reliability. Thus most farm trend and pressure indicators related to agricultural activity achieve a high score. The existing environmental data sets in the water, soil (and biodiversity) domains are far less developed in terms of coverage, time series and reliability. Consequently, the data required for pressure/State/impact indicators are often unavailable. Hence, several indicators of these domains have to rely on modelled or proxy data.

Where data quality appears high the spatial resolution of information can be disappointingly low. Data on nitrate concentration from Eurowaternet, for example, are only considered to be representative at EU-15 level.

Differences in data reliability and spatial resolution between indicators limit the possibilities for cross-referencing that is needed for a regional environmental analysis. This does not necessarily invalidate the potential usefulness of such indicators but shows that further effort is necessary in bringing together the more detailed data sets that are available at national level, e.g. for the monitoring of water quality. It is currently not feasible to fill the DPSIR framework for many storylines mainly due to the limited development of many indicators in the state and impact domains. The following section provides a more detailed review of data sets underpinning indicators in different environmental domains.

9.3 Review of data sets

Section 9.1 has reviewed the different functions of data sets for analytical purposes, which have become more demanding over time.

Section 9.2 provides an overview of the evaluation of the indicators, highlighting the increasing importance of spatial, integrated analysis that few existing data sets are designed for. This section reviews the individual data sets underpinning IRENA indicators, starting with agricultural and environmental data sources followed by modelling approaches and administrative data sets.

9.3.1 Review of agricultural data sources

Official statistics (FSS, FADN etc) are generally the most reliable and important data sets and are used in the following IRENA indicators: 'farmers' training levels' (IRENA 6), 'area under organic farming' (IRENA 7), 'water use (intensity)' (IRENA 10), 'energy use' (IRENA 11), 'cropping/livestock patterns' (IRENA 13), 'farm management practices' (soil cover, storage facilities for manure) (IRENA 14), 'intensification/extensification' (IRENA 15), 'specialisation/diversification' (IRENA 16), 'marginalisation' (IRENA 17), 'gross nitrogen balance' (IRENA 18), 'high nature value (farmland) areas' (IRENA 26), 'landscape state' (IRENA 32), 'impact on landscape diversity' (IRENA 35).

The combination of data sets and their validation through comparison still needs to be further explored. A good example would be the EU farm

Table 9.1 Evaluation overview

DPSIR	No	IRENA indicator	Range	Classification
Responses	1	Area under agri-environment support	13–15	Potentially useful*
	2	Regional levels of good farming practice	9–10	Potentially useful
	3	Regional levels of environmental targets	11	Potentially useful
	4	Area under nature protection	17	**Useful**
	5.1	Organic producer prices	13	Potentially useful
	5.2	Agricultural income of organic farmers	13	Potentially useful
	6	Farmers' training levels	13	Potentially useful
	7	Area under organic farming	18	**Useful**
Driving forces	8	Fertiliser consumption	14–15	Potentially useful/**Useful**
	9	Consumption of pesticides	12–14	Potentially useful
	10	Water use (intensity)	16	Potentially useful*
	11	Energy use	13–14	Potentially useful
	12	Land use change	15–17	**Useful**
	13	Cropping/livestock patterns	17–19	**Useful**
	14.1	Farm management practices- tillage	8	Potentially useful
	14.2	Farm management practices- soil cover	14	Potentially useful
	14.3	Farm management practices- manure	16	**Useful**
	15	Intensification/extensification	15	**Useful**
	16	Specialisation/diversification	15	**Useful**
	17	Marginalisation	13	Potentially useful
Pressures	18	Gross nitrogen balance	14	Potentially useful
	18sub	Atmospheric ammonia emissions	18	**Useful**
	19	Emissions of Methane (CH_4) and nitrous oxide (N_2O)	18	**Useful**
	20	Pesticide soil contamination	10	Potentially useful
	21	Use of sewage sludge	12	Potentially useful
	22	Water abstraction	11	Potentially useful
	23	Soil erosion	13	Potentially useful
	24	Land cover change	15–16	**Useful**
	25	Genetic diversity	12	Potentially useful
	26	High nature value farmland	12	Potentially useful
	27	Production of renewable energy (by source)	14	Potentially useful
State	28	Population of farmland birds	11–15	Potentially useful/**Useful**
	29	Soil quality	13	Potentially useful
	30.1	Nitrates in water	13	Potentially useful
	30.2	Pesticides in water	13	Potentially useful
	31	Ground water levels	6	**Low potential**
	32	Landscape state	12	Potentially useful
Impact	33	Impact on habitats and biodiversity	13	Potentially useful
	34.1	Share of agriculture in GHG emissions	19	**Useful**
	34.2	Share of agriculture in nitrate contamination	12	Potentially useful
	34.3	Share of agriculture in water use	9	Potentially useful
	35	Impact on landscape diversity	12	Potentially useful
Number of indicators	42		11 2 28 1	Useful Potentially useful/Useful Potentially useful Low potential

Note: A * behind the classification shows that the final classification of the indicator had to be downgraded due to insufficient score for the criteria relating to analytical soundness or data availability.

typology based on FSS and FADN data, which underpins the IRENA trend indicators (IRENA 13, 15 and 16).

There are other data sources which provide information related to agriculture: Land Use/Cover Area Frame Statistical Survey (LUCAS), OECD/Eurostat Joint questionnaire, DG Agriculture and Rural Development survey of organic farming.

9.3.1.1 Farm Structure Survey

The Farm Structure Survey (FSS) is constantly being reviewed with a view to adapting the survey to new user needs. In this process, it is sometimes decided to eliminate a number of less useful variables. In the context of agri-environment indicator development this requires an evaluation of the usefulness of individual variables from an environmental perspective.

When using FSS data it should be kept in mind that the main purpose of FSS is to follow structural trends in EU agriculture. Thus even the regular FSS censuses only includes holdings above a certain threshold and, for example, does not include common grazing land that is not allotted to individual holdings. Consequently, additional data sources will be useful when aiming to compare overall statistics on crop or livestock production between EU Member States.

- **Farmers' training levels (IRENA 6)** — FSS provides data on farmers' training levels in EU Member States. However, these variables do not report on training in environmentally friendly farming practices. Hence, at the moment the data available cannot be reliably linked to the relevant farm management behaviour of farmers, which makes it difficult to evaluate from an environmental perspective. It would appear useful, therefore, to review possibilities for including environmental training information into future FSS questionnaires.
- **Area under organic farming (IRENA 7)** — the FSS definitions on organic farming are based on Regulation (EEC) No 2092/91. Some Member States submit also areas receiving agri-environmental support for organic farming (for example in Sweden). IRENA 7 is, however, based on data that are submitted by Member States to DG AGRI in the framework of Regulation 2092/91 and include certified areas only. To avoid potential confusion it is recommended to assure that Member States adhere to the agreed FSS definitions.
- **Water use (intensity) (IRENA 10)** — variables related to irrigation are used as a proxy for water use intensity. The variables on type of irrigation equipment or techniques being used (surface, sprinkler, rain gun, drip) by holding introduced for the 2003 FSS survey can give an indication of how efficiently water is being used and should therefore be retained in future surveys.
- **Cropping/livestock patterns (IRENA 13)** — the analysis of changes in cropping patterns revealed that in some Member States there were large changes in utilised agricultural area (UAA) between 1990 and 2000 (see Section 3.3.1). In addition, FSS does not cover all agricultural land or livestock, as stated above. Hence other statistical data sets, such as land use or production statistics, need to be used in agri-environmental analysis when analysing crop and livestock patterns.
- **Farm management practices (soil cover, storage facilities for manure) (IRENA 14)** — crop area data is combined with expert knowledge to calculate the periods of the year that the soil is covered, which is important for limiting soil erosion (IRENA 23) and nutrient leaching (IRENA 18). Regional data on the area of spring and winter cereals should, therefore, be collected, though grain production statistics may be a better tool for this than FSS. The data on storage facilities for manure is important for reducing ammonia emissions and nutrient leaching. This FSS variable was first surveyed with the FSS 1993. From an environmental perspective it would be important to retain it as an obligatory variable in future FSS surveys and to ensure that the type of data collected is aligned with key environmental issues in manure management.
- **Specialisation/diversification (IRENA 16)** — the Community typology of farms is used for distinguishing between specialised and non-specialised farms. As stated previously, this joint typology is considered very useful and should be retained.
- **Gross nitrogen balance (IRENA 18)** — data on cropping area, livestock type and numbers, and nitrogen fixation crops (legumes and pulses) are used in combination with coefficients to calculate nutrient balances. FSS is the only survey that can provide important parameters for the calculation of regional nitrogen balances. The possibility of adding parameters to the FSS survey that are relevant for calculating such balances according to the OECD/Eurostat methodology should therefore be reviewed (as far as appropriate).

- **Landscape state (IRENA 32)** and **Impact on landscape diversity (IRENA 35)** — agricultural land use information is used to indicate the importance of agriculture in selected landscapes across Europe. No recommendation is put forward.

9.3.1.2 Farm accountancy data network (FADN)

The farm accountancy data network (FADN) consists of an annual survey carried out by the Member States of the EU, which collect accountancy data every year from a sample of the agricultural holdings. The main objective of FADN is the evaluation of the income of agricultural holdings and the analysis of economic impacts of the common agricultural policy. Derived from national surveys, FADN is the only harmonised micro-economic database, which combines data on farm structure, input use and economic variables. The combination of such different variables in one data set is a key factor for linking different issues in agri-environmental analysis. The data combination has been useful for developing the farm typology (Section 3.3.1.4), which is used to explain general trends in intensification/extensification (IRENA 15) and specialisation/diversification (IRENA 16). In addition, it is used to identify high nature value (farmland) areas (IRENA 26). Starting from the accounting year 2000, information on organic production methods is also collected. However, the sample of farms applying organic production methods is currently too small and this has hampered the use of the data for economic indicators on organic farming incomes (IRENA 5.2).

The FADN database only includes the 'commercial' farms beyond a certain economic threshold, which varies from one country to another according to the agricultural structure. This may lead to a certain under-representation of the smallest farms. In addition, FADN is only statistically representative at NUTS 0, 1 and 2 levels. FADN does not record the volumes of inputs used in specific production activities undertaken by the holding but only the total value of expenditure on certain inputs (such fertilisers, pesticides, feedstuff, energy, water, etc.) purchased by the holding (considered as a whole). The main recommendation is to broaden the survey to record the input volumes alongside the expenditure on inputs. In some Member States, this data is already available in the national FADN data sets. DG AGRI, responsible for managing FADN at EU level, carried out a survey (2002) on the availability of data concerning quantities of inputs ([45]). The results are:

- Energy — Three Member States (Belgium, Denmark and the Netherlands) have at least some information in their national FADN accounts returns, and Sweden was about to start collecting data for a sub-sample in 2002. Five Member States (France, Italy, Luxembourg, Finland, the United Kingdom) indicate that the data is available on the farm level, although only Italy and Finland do not set any conditions to include these data into the EU farm return.
- Fertilisers — Five Member States (Belgium, Ireland, Italy, Luxembourg and the Netherlands) already have information on the use of fertilisers in their national farm return, and Sweden was about to start collecting data for a sub-sample in 2002. Four Member States (Denmark, France, Finland, the United Kingdom) indicate that the data exist on the farm level, but they consider it very costly to include in the EU FADN farm return.

According to these results, it could be envisaged to include some environmental information (volumes of energy and fertilisers used) in FADN. In the case of water use, there are fewer possibilities as currently volumes of irrigation water are not consistently recorded at farm level.

9.3.1.3 Land use/cover area frame statistical survey (LUCAS)

The LUCAS survey has been used only in a limited way in IRENA due to its pilot character and the low sampling density. Its main contribution to IRENA is information on landscape features, which is used in 'landscape state' (IRENA 32). LUCAS transect data provide the number of agriculturally-linked linear elements per square kilometre for case study areas selected to illustrate the diversity of landscapes across Europe. Farm practice data was explored for inclusion in farm management practices (IRENA 14), but the data was not of sufficient quality to be included. LUCAS is a useful complementary tool as it provides geo-referenced land use and land cover data, which can help in validating Corine land cover. The short time span between data collection and availability is a strength of LUCAS. The usefulness of LUCAS would be further improved if a higher sampling density and accuracy could be achieved.

---

[45] Document RICC 1346 (2002) of the FADN Management Committee.

9.3.1.4 OECD/Eurostat Joint questionnaire

Information from the OECD/Eurostat Joint questionnaire has been used to underpin the water abstraction indicator (IRENA 22). Although it is supposed to be an annual survey, some Member States either repeat abstraction rates each year or the information is not provided. The joint questionnaire can only achieve its full potential if adequate cooperation is forthcoming from Member States. In this context, it would improve data interpretation if Member States could provide an explanation of the data provided (droughts, growth in irrigation area, new reservoirs etc.).

9.3.1.5 DG Agriculture and Rural Development survey of organic farming

Data is supplied by EU-15 Member States to DG Agriculture and Rural Development, using the administrative data from the organic farming questionnaire (electronic version OFIS). At present the completion of the organic farming questionnaire is partly voluntary, and there is a long time delay in the submission of data to DG Agriculture and Rural Development by some Member States. Given the dynamic development of this sector the annual reporting of data under the organic farming questionnaire appears important. For this instrument to be prioritised over FSS, however, reporting would have to become obligatory and time delays should be minimised.

9.3.2 Review of environmental data sets

Environmental data sets are used to monitor changes in land use and land cover; methane, nitrous oxide and ammonia emissions; nutrient levels in surface and ground water; pesticide levels in surface and ground water; farmland birds; important bird areas; and prime butterfly areas. They are often developed on the basis of reporting obligations arising from environmental legislation or international agreements (e.g. Nitrates Directive or the Kyoto protocol) but can also be primarily designed for monitoring environmental trends (e.g. Corine land cover or bird population data). Many of the biodiversity data sets used for IRENA indicators have the special common feature that they are collected by non-government organisations. The 'AROMIS database' was compiled in an EU-funded research project on the 'Assessment and reduction of heavy-metal input into agro-ecosystems' (AROMIS), but finally not utilised for the development of IRENA indicators.

The link between environmental reporting to environmental legislation has sometimes focused on the objectives and needs of that specific legislation without sufficient consideration for wider environmental reporting requirements. This factor, among others, may have contributed to the often-insufficient spatial resolution of environmental data sets that was encountered in the development of the IRENA indicators.

9.3.2.1 Corine land cover

Corine land cover 2000 (CLC 2000) is an update for the reference year 2000 of the first CLC inventory that was finalised in the early 1990s as part of the European Commission programme to COoRdinate Information on the Environment (Corine). It provides spatially referenced information on land cover and land cover changes during the past decade across Europe. CLC works on the principle of identifying land cover classes for polygons of a minimum size of 25 ha. This implies that it cannot provide correct land cover information for each individual land cover parcel but is representative over a wider area. Due to the spatial referencing of polygons it brings overall significant possibilities to environmental analysis, especially when combined with other data sets.

The ground observations collected by the LUCAS survey are already being used to validate CLC land cover classification, and similar opportunities should be explored. Work should also continue to ensure the full compatibility of 1990 and 2000 datasets. Improvements in the nomenclature could still be made to avoid potential confusion of terms with other environmental disciplines. For example, the term 'semi-natural' has a much wider meaning under Corine than in geo-botany from which it originates. Satellite-based information does not allow sufficient differentiation between different grassland classes from a biodiversity perspective. It should be explored if Corine can be complemented in this respect with ground-based grassland surveys that are available for at least some Member States.

9.3.2.2 European community greenhouse gas inventory 1990–2002 and inventory report 2004

Data for the emissions of methane and nitrous oxide (IRENA 19) and the share of agriculture in GHG emissions (IRENA 34.1) comes from the official national total and sectoral greenhouse gas emissions data submissions reported annually by Member States to UNFCCC, the EU monitoring mechanism and Eionet. The data is compiled for the EU by the EEA in the report (and related database) 'European community greenhouse gas inventory 1990–2002 and inventory report 2004', Technical report No 2/2004 (EEA, 2004a).

A recent workshop at the Joint Research Centre made both general and specific recommendations to provide better information on GHG emissions from agriculture (http://carbodat.ei.jrc.it/ccu/pweb/leip/home/ExpertMeetingCat4D/index.htm). The IRENA operation has used nationally reported data to underpin the indicators. It would, however, be more appropriate to use regional information. This would require the regionalisation of emission factors, as with the gross nutrient balance.

9.3.2.3 UNECE/EMEP Convention on Long-range Transboundary Atmospheric Pollution (CLRTAP)

Data for the atmospheric emissions of ammonia from agriculture (IRENA 18sub) is based on official national data submissions reported by Member States to the UNECE/EMEP Convention on Long-range Transboundary Atmospheric Pollution (CLRTAP). The IRENA operation has used national reported data to underpin the indicator. It would, however, be more appropriate to use regional information. These data feed into the calculation of gross nitrogen balance (IRENA 18), which should be reported at regional level. Moreover, it should be noted that the accuracy of data on the size of different emission sources (including the contribution of agriculture to air pollution) as well as emission coefficients could be improved. The EMEP programme (Cooperative Programme for Monitoring and Evaluation of the Long-range Transmission of Air pollutants in Europe) provides grid-level data on ammonia. EMEP uses monitoring sources and modelling techniques, which are in some cases different to the data provided under the UNECE protocol. However, grid-level data could also be used as a sub-indicator to provide a regional dimension.

9.3.2.4 Eurowaternet

Data for the indicators on nitrates and pesticides in water (IRENA 30) is based on information provided by Eurowaternet. Eurowaternet is a monitoring network designed for collecting data on the status and trends of Europe's water resources in terms of quality and quantity, and for analysing how this reflects pressures on the environment. In the future it will be adapted to meet the reporting needs of the water framework directive. Currently, Eurowaternet does not include enough monitoring stations to provide regional analysis. Data on nitrates in water is designed for reporting at EU-15 level, but IRENA 30 aggregates country information to southern, central and northern regions.

Pesticide information is only available from some national monitoring systems. Most of these only provide data for two pesticide compounds: simazine and atrazine.

Groundwater levels (IRENA 31) are not included in Eurowaternet at all, so this indicator is reliant on individual case studies.

Efforts are currently being made to geo-reference monitoring stations to the new catchment database developed by the Joint Research Centre. However, the monitoring stations included in Eurowaternet are not designed to monitor non-point sources of pollution from agriculture. Instead, stations are positioned to monitor major industries and sewage recycling plants, and may be moved after a few years. Therefore, a major investment is needed to meet the monitoring requirements of pollution from agriculture. The reporting needs of the water framework directive will go some way to improving the current data set on the state of nutrient and pesticide levels in surface and groundwater. The spatial distribution of this future monitoring network should aim to match the spatial resolution of data for related agriculture pressure indicators to enable effective agri-environmental analysis.

9.3.2.5 Pan-European common bird monitoring database

The Pan-European common bird monitoring database is maintained by the Royal Society for the Protection of Birds (RSPB), the European Bird Census Council (EBCC), and BirdLife International. The database is used to underpin the indicator on population trends of farmland birds (IRENA 28). Survey methods and data compilation follow tested and widely recognised approaches in the biological monitoring field. Data gathering is largely carried out by thousands of volunteer ornithologists who need to be trained appropriately to achieve maximum standardisation and data quality.

The set of 23 bird species associated with farmland habitats that are currently used to monitor the impact of agriculture on bird populations may be expanded or refined as further ecological information becomes available but is considered an appropriate selection by the experts consulted. As data becomes available, it would be useful to stratify population trends for individual bird species to the main agricultural habitats they occupy. This would allow a more detailed assessment of key agricultural land use trends on bird species by habitat and facilitate more targeted policy action where necessary.

9.3.2.6 Important bird areas

The database of important bird areas (IBAs) is maintained by BirdLife International. The network of IBAs has been identified on the basis of clear and consistent criteria that are applied across Europe and build on national bird distribution and bird population data. The database also provides information on land use or other threats to bird populations in IBAs, including agricultural intensification and agricultural abandonment (IRENA 33). The information to underpin the reported threats to IBAs is collected mostly by a network of volunteer or professional ornithologists that act as compilers of information for individual IBAs.

The criteria for distinguishing different land use and threat categories are published in the latest IBA publication. The reporting process is further harmonised via an electronic manual and reporting form that compilers are required to follow. BirdLife and partners also organise training seminars for IBA compilers and national coordinators. Quality checking procedures at central level and efforts undertaken to train compilers in reporting on threats to IBAs are documented. However, more detailed documentation of the quality assurance process and an assessment by country or IBA type would be useful.

9.3.2.7 Prime butterfly areas

The database of prime butterfly areas (PBAs) is maintained by De Vlinderstichting (Dutch Butterfly Conservation). PBAs are identified on the basis of standardised and reliable butterfly observation techniques and are likely to be a good minimum selection of areas important to butterfly conservation in Europe. The database also provides information on the threats to PBAs, including agricultural intensification and agricultural abandonment (IRENA 33). The methodology that is applied to distinguish between these two processes follows a standard procedure set out in a questionnaire (see van Sway & Warren, 2004).

Procedures to ensure a standardised treatment of this questionnaire between Member States and individual field workers need to be documented.

9.3.3 Review of modelling approaches

Modelling approaches are adopted for indicators where surveyed environmental data is not available. Models can be very useful tools for environmental analysis as long as the required input data are of sufficient quality. Quality input data are, however, not available for all models employed for IRENA indicators. In this case, the relevant indicators need to be regarded as an approximation. The modelled indicators are: 'gross nitrogen balance' (IRENA 18), 'pesticide soil contamination' (IRENA 20), 'soil erosion' (IRENA 23), and 'soil quality' (IRENA 29).

9.3.3.1 Gross nitrogen balance

The gross nutrient balance is a simple approach which combines nutrient inputs and outputs for estimating the levels of nutrient surpluses and deficits. At present Member States carry out their own national balances according to the agreed OECD/Eurostat methodology using standardised spreadsheets. For Member States that have not supplied nutrient balances, Farm Structure Survey data and coefficients from neighbouring Member States are used to determine national balances. It is known, however, that there are major regional differences within Member States. It would be more appropriate therefore to calculate regional balances. But, as in the case of GHG and ammonia emissions, regional balances will provide maximum information only in combination with regional coefficients (e.g. manure excretion rates, fertiliser application rates, and yields).

9.3.3.2 Pesticide soil contamination

The potential average annual content of herbicides in soils is modelled on the basis of herbicide degradation rates. The models rely on poor quality data on national application rates. No large scale measurement data are available to validate the annual estimates of annual herbicide content in the soil. Different regional application rates can be simulated, however, such data does not exist and/or is not available. It is recommended that more effort is made to obtain experimental data that can serve this purpose. It is further recommended to extent the list of herbicides used in crops under consideration so that information is not biased due to the limited number of herbicides addressed in the report. Once critical regions have been identified using the proposed indicator, more detailed analysis could be conducted. The next problem is thus to know what plant protection products farmers are actually applying, under which weather conditions and where in the catchment landscape. Such information can only be built up gradually on the basis of case study information, or regional/national surveys of farmer behaviour.

9.3.3.3 Pan-European soil erosion risk assessment (Pesera)

The Pan-European soil erosion risk assessment model (Gobin and Govers, 2003) is a process-based and spatially distributed model to quantify soil erosion by water and assess its risk across Europe. The model is intended as a regional diagnostic tool and can include scenario analysis for different land use and climate changes. Immediate improvements of the adopted Pesera model could include the use of newly available DTM data (90m resolution data from SRTM), the new Corine land cover 2000, more accurate rainfall data and more detailed soil information (1:250 000 scale). A crucial second step to improve the Pesera model is the use of land use and farm management data instead of land cover data. This should include crop rotations and agricultural practices applied (conventional tillage, reduced tillage, zero tillage, etc.). Only this could allow detecting impacts of agricultural policy reform.

9.3.3.4 Top soil organic matter estimates

Estimates of topsoil organic matter are used to underpin the soil quality indicator (IRENA 29). The use of soil organic matter as a proxy for soil quality requires further conceptual refinement. Threshold values need to be defined. EU soil organic carbon estimates are based on soil data (1:1 000 000 scale), temperature data and land cover from Corine land cover. Ground truth is needed with actual measurements in order to validate results. There is great potential here for the use of data collected directly by farmers through the regular analysis of their soils. The availability of this type of data is however hampered by privacy and copyright issues, but in those countries where these constraints have been overcome good results could be produced.

Immediate improvements of the estimates could be achieved through the use of more detailed soil data (1:250 000 scale) and the new Corine land cover 2000. In the longer term the real improvement would be achieved by using land use instead of land cover data. This could take into account farming practices, and therefore potentially detect impacts of agriculture policy reform. The LUCAS survey could be an appropriate tool to collect such data.

9.3.4 Review of administrative data sets

Administrative data is used to underpin the following response indicators: 'area under agri-environment support' (IRENA 1), 'regional levels of good farming practice' (IRENA 2) 'area under nature protection' (IRENA 4) as well as the sub-indicator related to agri-environmental training actions under rural development programmes (IRENA 6). The use of additional administrative data sets was explored but had to be discarded due to confidentiality rules with regard to access to national level source data. The most prominent examples of this issue are the reporting by Member States on nitrate concentrations in water bodies under the nitrates directive as well as livestock and parcel registers under the Integrated Administration Control System (IACS) that is used to manage CAP payments to farmers.

9.3.4.1 Common indicators for monitoring the implementation of rural development programmes

Member States have to report data annually to the European Commission on the area and expenditure of agri-environment measures under Regulation (EC) 1257/1999 according to pre-specified guidelines. However, the data supplied are not fully consistent between Member States with regard to the classification of agri-environment schemes by 'type of action'. The double counting of area in Member States where farmers can enrol the same land in two different agri-environment agreements is an additional difficulty in evaluating area coverage. By careful analysis of the data reported, this can largely be eliminated. Nevertheless, there is a need to further standardise the reporting by Member States to the Commission and to develop more coherent and clearly identifiable categories for agri-environment schemes targeted on different environmental issues.

The geo-referenced data on agri-environment scheme uptake that Member States have to provide (from 2005) through the Integrated Administration and Control System (IACS) will allow better spatial reporting. This should provide information not only on total area but also on the location of agri-environment contracts under different agri-environment schemes. This could allow assessing whether they are actually contributing to environmental objectives within already existing Community legislation (e.g. Natura 2000).

9.3.4.2 National/regional codes of good farming practice included in rural development programmes (RDPs)

The information required for this indicator does not come in a standard data format. Codes of GFP are farm management standards that farmers have to adhere to for being eligible for compensatory allowances in Less Favoured Areas and for joining agri-environment schemes. The codes of GFP set out in national and regional RDPs were used to compile an overview of GFP standards. Due to its qualitative nature, this information is not fully comparable between Member States. In addition, the indicator aims at understanding the extent to which national codes of GFP address key agri-environment issues. However, it is not a measure of the actual implementation of GFP by farmers as standards are not always compulsory. Reporting of Member States under the new cross compliance measure introduced with the CAP reform in 2003 may enable a better knowledge of farming practices in the future.

9.3.4.3 Natura 2000

Geo-referenced information on the extent and distribution of targeted agricultural habitat types at EU-15 level is not available. Instead data on geographical parameters and biological characteristics for candidate Natura 2000 sites were used, based on those reported by Member States in the standard data form for Natura 2000 (IRENA 4).

No common protocol exists for collecting data, and different approaches have therefore been adopted by Member States in filling out the standard data form. Some Member States use vegetation maps or surveys, whereas others may have used more intensive field studies when filling out the form. In addition, different habitat types may need different assessment techniques. Geo-referenced information on targeted agricultural habitats is essential for an environmental assessment.

The Natura 2000 network provides a chance for establishing standardised monitoring procedures for agricultural and other habitats. However, ground surveying and appropriate assessments are difficult to guarantee and thus require adequate support, guidelines and follow-up.

9.4 Conclusions

The IRENA operation has made an important contribution to developing agri-environmental indicators at EU-15 level. Many avenues in terms of indicator concepts and data sets have been explored — some more successful than others. The DPSIR framework remains a useful analytical framework for developing environmental storylines and is particularly helpful for explaining agri-environmental links. The environmental analysis presented in Chapters 4–8 has encountered clear limits. These lie on the one side in the limits of an indicator approach where contextual information based on research and knowledge of the agricultural sector is required to interpret and link indicator results. Secondly, the logic of the DPSIR framework was not always appropriate for the agri-environmental issue in question or could not be applied due to weaknesses in key indicators (e.g. water resources). Lastly, deficiencies in underpinning data sets in terms of data quality and/or geographic coverage are critical constraints. Differences in data reliability and spatial resolution between indicators limit the possibilities for cross-referencing that is needed for a regional environmental analysis.

The following sections summarise the main conclusions drawn from the evaluation of indicators and data sets and attempt to point to ways forward for future work.

9.4.1 Indicator evaluation

A quarter of the indicators scored 15 points or more implying that they are considered 'useful', 28 indicators scored between 8 to 14 points and were classified as 'potentially useful', and only 1 indicator was considered to be of 'Low potential' (ground water levels — IRENA 31) which means that further development would provide limited added value.

However, many indicators in the highest category show deficiencies in some key criteria, mainly due to a lack of time series data or spatial information. It has also proven difficult to link indicators from different data sets, usually because reporting levels are not consistent.

The air and climate change environmental storyline performed best for three reasons: firstly, the pressure and state/impact indicators are target-driven and developed to reach concrete goals; secondly, the storyline was developed using pressure and state indicators reported at national level, which is much less complicated than the regional approach adopted for the other analyses; and thirdly, the monitoring of changes in emissions is easier and cheaper than monitoring changes in biodiversity or water resources.

The agricultural water use environmental storyline performed the worst because quantitative information on irrigation and the impact on water supply was of low quality (IRENA 22 and 34.3) or unavailable (IRENA 31). Therefore, it was impossible to analyse the information using the DPSIR framework.

Overall, the scoring is strongly influenced by conceptual criteria, which shows that the original list of agri-environmental indicators was overall relevant. However, many indicators were downgraded from 'useful' to 'potentially useful' due to data difficulties, which demonstrates that a key challenge for indicator development is to build on or improve data sets.

Some of the challenges ahead for a better use of indicators in agri-environmental analysis are outlined below.

- **Regional information.** Information at NUTS 2/3 level (where available) is generally sufficient for describing agri-environmental patterns at EU-15 level, in particular for the driving force and pressure indicators developed during the IRENA operation. However, if we want to understand agri-environmental processes and causal links in sufficient detail for targeted policy action the level of spatial reporting of the state/impact/response indicators needs to be more detailed. For some issues, such as water quality, it may be necessary to develop indicators for catchments as well as administrative regions to strengthen the causal links between pressure (e.g. gross nitrogen balance) and pressure/state/impact indicators, which monitor catchments.
- **Precise spatial referencing** of relevant data sets in a geographical information system (GIS) is the key to improving environmental analysis. Only this allows for regional analysis and enables integration with other data sets.
- **Models.** In some cases, it may be more appropriate to adopt a modelling framework, especially if the state/impact indicators rely on modelled data. Modelling frameworks provide the possibility to evaluate the importance of input indicators using a sensitivity analysis. This may be more revealing than trying to link up different indicators within a DPSIR framework. However, at a European scale, it is difficult to obtain ground data to calibrate and validate estimates, and even the best modelling cannot improve inadequate input data.
- **Administrative data sets** can fill important gaps, but efforts should continue to improve such data sets in line with statistical and geo-referencing principles to obtain more added value. Geo-referenced farm registers seem a good way forward, in particular if further attributes relevant to rural development issues and environmental analysis, such as input use or manure systems, can be added. Administrative data are by their nature, however, not as stable as official statistical data sets.
- **Integration of data bases.** The IRENA operation has used a number of different data sets to develop indicators. There is a need to integrate such data sets to achieve added value and common analytical objectives. For example, the integration of LUCAS (ground observations) and Corine land cover (satellite image interpretation) may enable improvements to the validation of Corine land cover information. Farm structure survey data may also improve the information on changes in agricultural land uses in Corine land cover.
- **Spatial modelling.** Spatialisation methods offer further opportunities although this technique needs further development and validation. The redistribution of pressure indicators derived from farm census data, reported at administrative level, to catchments can be done by spatial modelling on the basis of Corine land cover variables (Campling et al., 2005).

9.4.2 Operational issues for improving the quality of data sets and user access

If time-series data are to be useful for environmental analysis, they have to be comparable between years. For instance, a change in threshold values for FSS data collection since 1990 has diminished the comparability of 1990 and 2000 data. Where such threshold changes are necessary it would make data time-series analysis much easier if data prior to the threshold change could be adapted to the new definition.

Corine land cover data is often the only option for estimating spatial distribution of livestock and land use data, (Campling et al., 2005) even though Farm Structure Survey data is collected at NUTS 4/5 levels. Making farm census data available at NUTS 4/5 levels would considerably improve indicator information, but will require a change of confidentiality rules. This would make the integration of spatial data sets belonging to different components of the DPSIR framework much easier and would reduce uncertainty.

Non-government organisations are carrying out important work in providing valuable state/impact data concerning farmland birds and butterflies. This data needs to be better analysed in the context of the farm trend indicators established by IRENA, so that possible quantitative links can be established. It also appears important to ensure the long-term viability of this data collection exercise that is currently funded largely through private donations of time and money. Options for ensuring such continuity need to be considered at EU level.

9.4.3 Future perspectives and challenges

Society's expectations of agriculture have evolved from food production to 'multi-functional services' with an important role for environmental management. Following society's concern, the environmental considerations play an important role in the common agricultural policy. Thus, relevant statistical data in the domain of rural development and the environment become increasingly important, and agricultural statistics have to serve wider purposes than analysing production data and farm trends. Consequently, an agri-environmental information system has to be developed that allows policy makers to build the right framework for the farming sector to fulfil these expectations. The results of the IRENA operation ([46]) form a good basis from which the building of such an agri-environmental information system can continue.

There are many challenges ahead in terms of improving data sets, spatial referencing and ensuring the timely delivery of indicators to policy makers. It is important that the current list of indicators is reviewed and, if necessary, amended to meet current analytical and monitoring needs. This includes deciding which reporting scale is strictly necessary at the EU-15 level, especially in light of the current deficiencies in existing data sets highlighted in this report. At the scale feasible for maps in this report it has proven difficult to show sufficient detail for many indicators compiled at regional level. The need to extend indicator-based reporting to include new and future EU Member States has to be taken into account in this context. Resource limitations at national and EU level probably make it necessary to limit future agri-environment indicator work to a reduced set. Experiences gained on technical possibilities and a careful evaluation of policy relevance should be the guiding criteria in this regard.

Reporting scale is an important determinant of database and indicator development. Data sets for reporting at EU-level can be coarser than those for national or regional analysis. However, EU indicator data sets are ideally aggregated from more local, spatial information. Data sets should thus be nested (i.e. build on each other), which also allows the further detailed analysis of agri-environmental issues that at EU-level can only be identified but not analysed.

The farm typology approach could be further explored as a means to relate indicators to different agricultural sectors, and to integrate this information with other indicators. This would facilitate the interpretation of indicator results and allows decision makers to focus in on particular farm types. To develop operational indicators, a consistent reporting scale needs to be adopted; otherwise important information gaps may emerge. Interoperability between different data sets is particularly important in this field, for example between the Farm Structure Survey (FSS) and the farm accountancy data network (FADN).

There are other initiatives concerned with developing and reporting on European wide data sets, such as Global Monitoring for Environment and Security (GMES) and Infrastructure for spatial information in Europe (INSPIRE). These initiatives aim to develop a harmonised and standardised platform of spatial data, which future agri-environmental indicator development could exploit. Cooperation with these large initiatives requires effective communication between all relevant organisations at EU and national level. Cooperation and communication in agri-environmental indicator work has been one of the main achievements of the IRENA operation, which should continue in the future.

([46]) These are the present indicator report, the individual indicator fact sheets and the underlying data bases and an indicator-based assessment report on the integration of environmental concerns into the CAP.

List of acronyms

AEI	Agri-environment indicator
CAFE	Clean Air for Europe
CAP	Common agricultural policy of the European Union
CH_4	Methane
CLC	Corine land cover (a land cover survey using satellite images)
CO_2	Carbon dioxide
DAD-IS	Domestic Animal Diversity Information System (maintained by FAO)
DG AGRI	Directorate-General for Agriculture and Rural Development
DG ENV	Directorate-General for Environment, Nuclear safety and Civil protection
Dismed	Desertification information system project for the Mediterranean
DPSIR	Driving Forces-Pressures-State-Impact-Responses (a framework for environmental analysis and indicator classification)
EAGGF	European Agriculture Guarantee and Guidance Fund
EBCC	European Bird Census Council
ECPA	European Crop Protection Association
EEA	European Environment Agency
EFMA	European Fertiliser Manufacturers Association
Eionet	European environment information and observation network
EMEP	Cooperative Program for Monitoring and Evaluation of the Long-range transmissions of air Pollutants in Europe (under the Convention on Long-range Transboundary Air Pollution)
ESB	European Soil Bureau
ETC	European topic centre
Eurostat	Statistical Office of the European Communities
EU	European Union
EU-12	Belgium, Denmark, France, Germany, Greece, Ireland, Italy, Luxemburg, Portugal, Spain, the Netherlands, the United Kingdom
EU-15	Austria, Belgium, Denmark, Finland, France, Germany, Greece, Ireland, Italy, Luxemburg, Portugal, Spain, Sweden, the Netherlands, the United Kingdom
ESDP	European Spatial Development Perspective
FADN	Farm accountancy data network — (a sample survey used to gain insight into the economic situation of farms in the EU)
FAO	Food and Agriculture Organization of the United Nations
Faostat	On-line and multilingual database covering international statistics in several areas related to agriculture (FAO)
FOCUS	Forum for the coordination of pesticide fate models and their use
FSS	Farm Structure Survey — (the main statistical survey of farms and agricultural land use in the EU)
GFP	Good farming practice
GHG	Greenhouse gases
GJ	Giga Joule (Giga is 109, and one joule is the equivalent of one watt of power radiated or dissipated for one second)
GIS	Geographic Information System
GLS	Grazing Livestock System
GTOPo30	Global Digital Elevation Model (which translates altitude information into a GIS system)
GWP	Global Warming Potential (a term used to compare the potential climate change impact of different greenhouse gases)
ha	Hectare
HAIR	Harmonised pesticide risk indicators
HARM region	Regional division, which gives the opportunity to compare Farm Structure Survey NUTS 2 regions with farm accountancy data network regions
HNV	High Nature Value (farmland)
IBA	Important Bird Area (identified via bird population data, compiled by BirdLife International and partners)
IFEN	L'Institut Français de l'ENvironnement
IFOAM	International Federation of Organic Agriculture Movements
INEA	Istituto Nazionale de Economia Agraria
IPCC	Intergovernmental Panel on Climate Change
IRENA	Indicator Reporting on the Integration of Environmental Concerns into Agricultural Policy
JRC	Joint Research Centre of the European Commission
kt	Kilo tonnes
LEAC	Land and Ecosystems Account

List of acronyms

LEI	Landbouw Economisch Instituut
LiM	Landscape inventory Method
LFA	Less favoured area
LU	Livestock unit
LUCAS	Land Use/Cover Area Frame Statistical Survey
MARS	Monitoring Agriculture by Remote Sensing (a programme run by the Joint Research Centre)
N	Nitrogen
NEC	National Emissions Ceilings
NO_3	Nitrates
N_2O	Nitrous oxide
NH_3	Ammonia
NUTS	Nomenclature of Territorial Units for Statistics
NVZ	Nitrate Vulnerable Zone
OECD	Organisation for Economic Cooperation and Development
OMIaRD	Organic marketing initiatives and rural development (an EU-funded research project)
P	Phosphorus
PAIS	Proposal on Agri-Environmental Indicators (a project financed by Eurostat)
PBA	Prime Butterfly Area
PECBM	Pan-European Common Bird Monitoring project (run by EBCC, BirdLife International and national partners)
Pesera	Pan-European soil erosion risk assessment (a research project)
PPPs	Plant Protection Products
Ramsar	Convention on cooperation for the conservation and wise use of wetlands and their resources, signed in Ramsar, Iran, in 1971
RSPB	Royal Society for the Protection of Birds
RDP	Rural Development Programme
TOE	Tonnes of oil equivalent (the energy content of one tonne of oil)
TRIM	TRends and Indices for Monitoring data (a statistical model)
UAA	Utilised Agricultural Area
UNECE	United Nations Economic Commission for Europe
UNFCCC	United Nations Framework Convention on Climate Change
WTO	World Trade Organisation

References

Agriculture Council (1999). *Strategy on environmental integration and sustainable development in the common agricultural policy established by the Agriculture Council.* Document submitted to the European Council In Helsinki, 13078/99. http://ue.eu.int/ueDocs/cms_Data/docs/pressdata/en/misc/13078.en9.htm.

Amann, M., Bertok, I., Cofala, J, Gyarfas, F., Heyes, C., Klimont, Z., Schopp, W. and Winiwarter, W. (2005). *Baseline scenarios for the Clean Air for Europe (CAFÉ) programme.* International Institute for Applied Systems Analysis. http://www.iiasa.ac.at/rains/CAFE_files/Cafe-Lot1_FINAL(Oct).pdf.

Baldock, D., Beaufoy, G., Brouwer and Godeschalk, F. (1996). *Farming at the Margins, Abandonment or redeployment of agricultural land in Europe.* IEEP and LEI-DLO: London and The Hague.

Baldock, D., Caraveli, H., Dwyer, J., Einschütz, S., Petersen, J.E., Sumpsi-Vinas, J. and Varela-Ortega, C. (2000). *The environmental impacts of irrigation in Europe. A report to the Environment Directorate of the European Commission.* Institute for European Environmental Policy in association with the Polytechnical University of Madrid and the University of Athens.

Bignal, E.M. and McCracken, D.I. (1996). Low-intensity farming systems in the conservation of the countryside. *Journal of Applied Ecology* 33: 413–424.

BirdLife International (2004). *Birds in the European Union: a status assessment.* Wageningen, the Netherlands, BirdLife International.

BirdLife International/EBCC (2000). *European Bird Populations — Estimates and trends.* BirdLife International Conservation series No 10.

Blum, W.E.H. and Varallyay, G. (2004). *Soil indicators and their practical application, bridging between science, politics and decision-making.* EUROSOIL 2004 4–12th September, 2004, Freiburg, Germany.

Brouwer, F., Baldock D. and La Chapelle, D., (eds) (2001). *High level conference on EU enlargement: the relation between agriculture and nature management*, Wassenaar, 22–24 January 2001.

Campbell, L.H. and Cooke, A.S. (eds.) (1997). *The indirect effects of pesticides on birds.* Peterborough, Joint Nature Conservation Committee, United Kingdom.

Campling, P., Terres, J.M., Vandewalle, S., Van Orshoven, J. and Crouzet, P. (2005). Calculation of Agricultural Nitrogen Quantity for EU15, spatialisation of the results to river basins using Corine land cover. *Physics and Chemistry of the Earth.* Volume 30, Issues 1–3, 2005, Pages 25–34, Elsevier, the Netherlands.

Countryside Survey (2000). *Accounting for Nature: Assessing Habitats in the UK Countryside,.*Department for Environment, Food and Rural Affairs http://www.defra.gov.uk/wildlife-countryside/cs2000.

De Angelis, A. (2002). *Towards a sustainable agriculture and rural development: agri-environmental indicators as elements of an information system for policy evaluation.* ARIADNE International Conference, Chania/Crete, Greece. 13–15 November 2002. Mediterranean Agronomic Institute of Chania (MAICh). http://www.ariadne2002.gr.

DISMED (2005). *Desertification information system project for the Mediterranean.* http://dismed.eionet.eu.int/index_html.

Donald, P.F., Green, R.E, and Heath, M.F. (2001). Agricultural intensification and the collapse of Europe's farmland bird populations. *Proceedings of the Royal Society Lond*on, 268, pp. 25–29.

EEA (1999). *Environmental indicators: typology and overview.* Technical report No 25. European Environment Agency, Copenhagen, Denmark.

EEA (2003a). *Europe's water: An indicator-based assessment.* Environmental issue report No 34. European Environment Agency, Copenhagen, Denmark.

EEA (2003b). *Assessment and reporting on soil erosion.* Background and workshop report. European Environment Agency, Copenhagen, Denmark.

EEA (2004a). *Greenhouse gas emission trends and projections in Europe 2004.* EEA Report No 5/2004.

European Environment Agency, Copenhagen, Denmark.

EEA (2004b). *Greenhouse gas emission trends and projections in Europe 2003*. Tracking progress by the EU and acceding and candidate countries towards achieving their Kyoto Protocol targets. European Environment Agency, Copenhagen, Denmark.

EEA (2004c). *High nature value farmland — characteristics, trends and policy challenges*. European Environment Agency, Copenhagen, Denmark.

EEA data service (2004). (http://dataservice.eea.eu.int/dataservice/available2.asp?type=findtheme&theme=water).

English Nature (2003). *England's best wildlife and geological sites: The condition of Sites of Special Scientific Interest in England in 2003*. English Nature. Peterborough, United Kingdom.

Environment Agency of England and Wales, (2003). Pesticides in rivers, groundwater and pollution incidents. http://www.environment-agency.gov.uk/yourenv/eff/business_industry/agri/pests/915588/?version=1&lang=_e.

European Commission (2000). *Communication from the Commission to the Council and the European Parliament, Indicators for the Integration of Environmental Concerns into the common agricultural policy*, COM (2000) 20 final.

European Commission (2000). *Communication from the Commission to the Council, the European Parliament, the Economic and Social Committee and the Committee of the regions: towards a thematic strategy for soil protection*, COM (2002) 179 final.

European Commission (2001). *Communication from the Commission to the Council and the European Parliament, Statistical Information needed for Indicators to monitor the Integration of Environmental Concerns into the common agricultural policy*, COM (2001) 144 final.

European Commission (2002). *Implementation of Council Directive 91/676/EEC concerning the protection of waters against pollution caused by nitrates from agricultural sources*. Synthesis from year 2000 Member States reports. http://europa.eu.int/smartapi/cgi/sga_doc?smartapi!celexplus!prod!DocNumber&lg=en&type_doc=COMfinal&an_doc=2002&nu_doc=407.

European Council (2005) *Presidency Conclusions*. 22 and 23 March 2005. www.eu2005.lu/en/actualites/conseil/2005/03/23conseileuropen/ceconcl.pdf

Eurostat (2004). *Eurostat's concepts and definitions database*. Glossary: Livestock Unit (LU). http://forum.europa.eu.int/irc/dsis/coded/info/data/coded/en/gl009931.htm.

GEUS (2002). *Grundsvandsovervagning 2002*. Danmarks og Groenlands geologiske undersoegelser. Ministry of Environment, Copenhagen.

Gobin, A. and Govers, G. (2003). *Pan-European soil erosion risk assessment Project* (Pesera). Third annual report to the European Commission. EC Contract No QLK5-CT-1999-01323.

Gobin, A., Jones, R. Kirkby, M., Campling, P., Govers, G., Kosmas, C., and Gentile A.R. (2004). 'Indicators for pan-European assessment and monitoring of soil erosion by water'. *Environmental Science and Policy* 7: 25–38. Elsevier, the Netherlands.

Heath, M. F. and Evans, M. L. (2000). *important bird areas in Europe: Priority sites for conservation — Volume 1 and 2: Northern Europe*. BirdLife Conservation Series No 8. Cambridge, BirdLife International.

Hole, D.G., Perkins, A.J., Wilson, J.D., Alexander, I.H., Grice, P.V and Evans, A.D. (2005). 'Does Organic farming benefit biodiversity?' *Biological Conservation* 122 (2005) 113–130.

IFEN (2004) Les pesticides dans les eaux, 6ème bilan annuel. Données 2002/Pesticides in water. 6th annual report. 2002 data (http://www.ifen.fr/publications/ET/pdf/et42.pdf).

Jørgensen, U. and Kristensen, E.S. (2003). *Organic farming benefits the aquatic environment*. Newsletter from Danish Research Centre for Organic Farming. December 2003, No 4.

McCubbin, D.R., Apelberg, B.J., Roe, S., and Divita F. Jr (2002) Livestock ammonia management and particulate-related health benefits. *Environmental Science and Technology*, 15;36 (6):1141-6.

OECD (1993). *OECD core set of indicators for environmental performance reviews: a synthesis report by the group on the state of the environment*. Environment Monographs 83, Organisation for Economic Co-operation and Development, Paris.

OECD (2006, forthcoming). *Environmental Indicators for Agriculture Volume 4*. OECD publications service,

References

Organisation for Economic Co-operation and Development, Paris. http://www.oecd.org/agr/env/indicators.htm.

Pain, D.J. and Pienkowski, M.W. (eds) (1996). *Farming and birds in Europe: the common agricultural policy and its implications for bird conservation.*. Academic Press, London.

PAIS II (2005).*Proposal on Agri-environmental Indicators*. Report prepared by Landsis g.e.i.e, Luxembourg, for the European Commission — Eurostat.

Pannekoek, J. and van Strien, A. (1998). *TRIM 2.0 for Windows* (Trends & Indices for Monitoring data). Statistics Netherlands, Voorburg.

Potts, G.R. (1986). *The Partridge: Pesticides, Predation and Conservation*. Collins, London.

Shepherd, M., Pearce, B., Cormack, B., Philipps, L., Cuttle, S., Bhogal, A., Costigan, P., and Unwin, R. (2003). *An assessment of the environmental impacts of organic farming*. DEFRA, ADAS, ELM FARM, and IGER.

Souchère, V, King, B., Dubreuil, N., Lecomte-Morel V., Le Bissonnais, Y. Chalat, M. (2003). 'Grassland and crop trends: role of the European Union common agricultural policy and consequences for runoff and soil erosion'. *Environmental Science & Policy*. Volume 6, Issue 1, February 2003, pp. 7–16.

Spanish Ministry of Environment (2000). *Libro Blanco del Agua en España*, Madrid. http://hispagua.cedex.es/documentacion/documentos/l_b/l_b.php?localizacion=Libro %20Blanco %20del %20Agua.

Stolze, M.,Piorr, A.,Häring, A.M. and Dabbert, S. (2000). The environmental impacts of organic farming in Europe. *Organic Farming in Europe: Economics and Policy* Vol. 6. Universität Hohenheim, Stuttgart-Hohenheim.

Stoate, C., Boatman, N.D., Barralho, R.J, Rio Carvalho, C., de Snoo, and G.R.,Eden, P. (2001). Ecological impacts of arable intensification in Europe. *Journal of Environmental Management*, 63, 337–365.

Sumpsi, J, Blanco, M., and Varela-Ortega, C. (2000). *Politicas agrarias alternatives para reducir el uso del agua en los regadios de Daimiel*. Working paper, Research report for ADENA-WWF, Department of Agricultural Economics, Polytechnical University of Madrid.

Swaay, C.A.M. van and Warren, M.S. (eds.) (2003). *Prime butterfly areas in Europe: Priority sites for conservation*. National Reference Centre, Ministry of Agriculture, Nature and Food Quality, the Netherlands.

Swedish EPA (2002): http://www.internat.naturvardsverket.se/.

UBA (Federal Environment Agency) Vienna (2001). *Umweltsituation in Österreich*. Umweltbundesamt, Wien.

Vickery, J.A., J.R. Tallowin, R.E. Feber, E.J. Asteraki, P.W. Atkinson, R.J. Fuller and V.K. Brown. (2001). The management of lowland neutral grasslands in Britain: effects of agricultural practices on birds and their food resources. *Journal of Applied Ecology* 38:647–664.

Legislation referred to in the text

Council Directive 79/409/EEC of 2 April 1979 on the conservation of wild birds (OJ L103, 25.4.1979).

Council Directive 86/278/EEC of 12 June 1986 on the protection of the environment, and in particular of the soil, when sewage sludge is used in agriculture (OJ L181, 4.7.1986).

Council Directive 91/676/EEC of 12 December 1991 concerning the protection of waters against pollution caused by nitrates from agricultural sources (OJ L375, 31.12.1991).

Council Directive 91/414/EEC of 15 July 1991 concerning the placing of plant protection products on the market (OJ L230, 19.8.1991).

Council Directive 92/43/EEC of 21 May 1992 on the conservation of natural habitats and of wildlife and flora and fauna (OJ L206, 22.7.1992).

Council Directive 01/81/EEC of 23 October 2001 on national emission ceilings for certain atmospheric pollutants.

Council Decision 02/358/EEC of 25 April 2002 concerning the approval, on behalf of the European Community, of the Kyoto Protocol to the United Nations Framework Convention on Climate Change and the joint fulfilment of commitments there under.

Council Decision 02/1600/EEC of the European Parliament and of the Council of 22 July 2002 laying down the Sixth Community environment action programme (OJ L 242, 10/09/2002, p. 1).

Annexes

Annex 1

Table A.1 Changes in indicator titles

No	Original title	Proposed title
5.1	Organic producer prices	Organic producer prices and market share
5.2	Agricultural income of organic farmers	Organic farm incomes
8	Fertiliser consumption	Mineral fertiliser consumption
12	Land use: topological change	Land use change
14	Management practices	Farm management practices
18	Soil surface nutrient balance	Gross nitrogen balance
18sub [47]		Atmospheric emissions of ammonia
19	Methane emissions	Emissions of methane and nitrous oxide
21	Water contamination	Use of sewage sludge
22	(Ground) water abstraction	Water abstraction
26	High nature value (farming) areas	High nature value (farmland) areas
28	Species richness	Population trends of farmland birds

---

[47] 18sub 'Atmospheric emissions of ammonia from agriculture' is a new indicator proposed during the IRENA operation.

Annexes

Annex 2

Three farm typologies have been developed in the IRENA operation to help characterise general regional trends. These are required to reflect the different dimensions (input use, farm system, specialisation) that need to be explored in a farm trend analysis. The first typology (related to intensification/extensification) differentiates farms according to the expenditure on purchased farm inputs, using data from FADN. Expenditure is regarded as a proxy for input use. The classification of farm types into low-, medium- and high-input systems is based on the expenditure on fertiliser, crop protection and purchased concentrated feedstuff per ha per year (Table A.2).

The second typology differentiates farms based both on the Community Typology of agricultural holdings and land use criteria, using data from FADN to differentiate holdings according to their type of farming (e.g. grazing livestock, cropping — specialist crops, horticulture etc.).

A third typology is used for the specialisation/diversification indicator, which groups the Community Typology farm types into specialised and non-specialised categories.

Table A.2   IRENA farm type based on expenditure on fertiliser, crop protection and concentrated feedstuff per ha per year

IRENA farm type	Expenditure threshold (Euro/ha/year)
Low-input	< 80 EUR
Medium-input	80–250 EUR
High-input	> 250 EUR

Table A.3   IRENA farm type based on the Community typology and certain land use criteria

IRENA farm type	Community typology	Other criteria
Grazing livestock _Permanent Grass	4	> = 55 % of UAA in grass and < 40 % of grass in temporary grass
Grazing livestock _Temporary Grass	4	> = 55 % of UAA in grass and > = 40 % of grass in temporary grass
Grazing livestock _Forage Crops	4	Not grazing livestock_permanent grass or grazing livestock_forage crops
Pigs-poultry	5	
Cropping_fallow land	1+6	< 55 % of UAA in grass and >= 12.5 % of UAA in fallow)
Cropping_cereals	1+6	< 55 % of UAA in grass and < 12.5 % of UAA in fallow and >= 55 % of UAA in cereals)
Cropping_specialist crops	1+6	< 55 % of UAA in grass and < 12.5 % of UAA in fallow and < 55 % of UAA in cereals and > = 25 % of arable land in specialised crops (sugar beet, oil seed, seeds for sowing, potato, cotton and tobacco)
Cropping_mixed crops	1+6	Not cropping cereals, cropping specialist crops or cropping_fallow land
Horticulture	2	
Permanent crops	3	
Mixed cropping-livestock	7+8	

Table A.4   IRENA farm types based on the Community typology grouped in specialised and non-specialised categories

IRENA farm type	Community typology code	Community typology name
Specialised cropping	1	Specialist field crops
	2	Specialist horticulture
	3	Specialist permanent crops
Specialised livestock	41	Specialist dairying
	42	Specialist cattle-rearing and fattening
	441	Specialist sheep
	443	Specialist goats
	501	Specialist pigs
	502	Specialist poultry
Non-specialised livestock	442	Sheep and cattle combined
	444	Various grazing livestock
	503	Various granivores combined
	7	Mixed livestock holdings
Non-specialised cropping	6	Mixed cropping
Non-specialised cropping/livestock	8	Mixed crops-livestock

Annex 3

Table A.5 Development of agri-environmental indicators in the IRENA operation ([48])

Domain/sub-domain	No	IRENA indicator	Headline indicator and sub-indicators	Data sources	Spatial scale	Temporal scale
Responses: Public policy	1	Area under agri-environment support	Trends in the agricultural area enrolled in agri-environmental measures and share of the total agricultural area.	Common indicators for monitoring the implementation of RDPs, DG AGRI.	NUTS 0 / rural development programming regions	1998–2002
			1) Trends in agri-environment expenditure per hectare of utilised agricultural area (UAA)	1) European Agriculture Guarantee and Guidance Fund (EAGGF), DG AGRI.	NUTS 0 level	1) 2000–2003
			2) Endangered breeds under agri-environment measures.	2) Common indicators for monitoring of implementation of RDPs, DG AGRI.		2) 2001
	2	Regional levels of good farming practice	Range and type of relevant categories of farming practices covered by the codes of good farming practices defined by regions in their rural development programmes.	National/regional codes of good farming practices included in rural development programmes (RDPs) (period 2000–2006)	NUTS 0 level, except Belgium (2 = NUTS 1) and Italy (1 = NUTS 2 region)	Current status in 2004
			1) The 'regulatory' (requirements based on legislation) or 'advisory' approach (based on recommendations) taken by Member States in preparing their code of GFP.			
			2) The range of GFP requirements being verifiable standards (subject to control).			
	3	Regional levels of environmental targets	Environmental targets set at Member State level relevant to agriculture.	Commission and national policy documents	NUTS 0	Current status in 2004
	4	Area under nature protection	Proportion of Natura 2000 sites covered by targeted habitats that depend on a continuation of extensive farming practices.	Database of sites proposed under the habitats directive as NATURA 2000 areas	NUTS 0	Data received between 1997 and March 2005
					NUTS 2 and 3	Data received by July 2004
Responses: Market signals	5.1	Organic producer prices and market share	Organic producer prices and market share (to indicate levels of consumer demand for organic products and market signals to organic producers).	Research project OMIaRD (Organic marketing initiatives and rural development)	NUTS 0	2000, 2001
	5.2	Organic farm incomes	Organic farm incomes compared to similar conventional farms (to indicate combined impacts of prices, agri-environmental support payments and other factors on financial viability of organic holdings).	FADN	NUTS 0	Partial coverage 2000, complete coverage 2001
Technology skills	6	Farmers' training levels	The level of agricultural training of managers of agricultural holdings.	FSS	NUTS 2 and 3	1990–2000
			Training in agri-environmental issues.	Common indicators for monitoring the implementation of RDPs, DG AGRI.	NUTS 0	2001
Attitudes	7	Area under organic farming	Trends in organic farming area and in the share of organic farming area in the total utilised agricultural area (UAA).	Organic farming questionnaire on Regulation No 2092/91 (EEC) 1998–2002, DG AGRI; and FSS for regional share	NUTS 0	1998–2002
					NUTS 2 and 3	2000

---

[48] The acronyms used are: **CLC** (Corine land cover), **ECPA** (European Crop Protection Association), **EFMA** (European Fertiliser Manufacturers Association), **FSS** (Farm Structure Survey), **FADN** (Farm accountancy data network), **RDP** (Rural Development Programme), **SIRENE** (section of the Eurostat-New Cronos database with information on energy use in agriculture).

Annexes

Domain/ sub-domain	No	IRENA indicator	Headline indicator and *sub-indicators*	Data sources	Spatial scale	Temporal scale
Driving forces: Input use	8	Mineral fertiliser consumption	Mineral fertiliser consumption is indicated by the evolution of the consumption of nitrogenous (N) and phosphate (P_2O_3) mineral fertilisers over time.	Faostat	NUTS 0	Most recent 2002 Trend 1990–2001
			Fertiliser application rates for selected crops.	EFMA	NUTS 0	Most recent 1999/2000
	9	Consumption of pesticides	The consumption of pesticides (here plant protection products, excluding biocides and disinfectant products) is indicated by: (a) Used/sold quantities of different pesticide categories; (b) Application rates of different pesticide categories (insecticides/herbicides/others).	ECPA (use data) Member States (sales data)	NUTS 0	Use: 1992–1999 Sales: 1992–2002
	10	Water use (intensity)	a) Trend in irrigable area (area covered with irrigation infrastructure) and b) trends in total area (and by crops) irrigated at least once a year (actual area irrigated).	FSS	NUTS 2 and 3 (only Greece, France, Spain reported b) in 1990–2000)	Most recent 2000 Trend 1990–2000
			Trend of share of irrigable area in total UAA.	FSS	NUTS 2 and 3	Most recent 2000 Trend 1990–2000
	11	Energy use	Energy use is indicated by the annual use of energy at farm level by fuel type (GJ/ha).	FADN, SIRENE, FSS	NUTS 0 (and 1)	Trend 1990–2000
			Estimate of energy used to produce mineral fertilisers for agricultural use (GJ/ha).	Faostat for fertiliser use, 'energy content' based on industry data (the Netherlands)	NUTS 0	Trend 1990–2000
Driving forces: Land use	12	Land use change	Area of land use change from agriculture to artificial surfaces between 1990 and 2000.	CLC 1990 and 2000	NUTS 2 and 3	1990–2000
			Sector share of land converted from agriculture to artificial surfaces.	CLC 1990 and 2000	NUTS 2 and 3	1990–2000
	13	Cropping/ livestock patterns	Cropping patterns: trends in the share of the utilised agricultural area occupied by the major agricultural land uses (arable, permanent grassland and permanent crops). Livestock patterns: trends in the share of major livestock types (cattle, sheep and pigs).	FSS, FADN	FSS: NUTS 2 and 3 FADN: NUTS 0 and 1	1990–2000
			Trends of types of farms particularly relevant for environment (typology).			
Farm management	14	Farm management practices	1) Cropping methods: soil cover.	FSS	NUTS 2 and 3	2000
			2) Cropping methods: tillage method.	PAIS II project (2005)	NUTS 0	Only 2003/2004
			3) Type and capacity of storage for farm manure and slurry.	FSS	NUTS 2 and 3	2000
Driving forces: Trends	15	Intensification/ extensification	a) Trends in the share of agricultural area managed by low-input, medium-input or high-input farm types (based on the average expenditure on inputs per hectare).	FADN	FADN: NUTS 0 and 1	1990 and 2000
			b) Livestock stocking densities.	FSS, FADN.	FSS: NUTS 2 and 3 FADN: NUTS 0 and 1	1990 and 2000
			c) Trends in yields of milk and cereals.	FADN	FADN: NUTS 0 and 1	1990, 1997, 2000
	16	Specialisation/ diversification	Specialisation is indicated by trends in the share of the agricultural area managed by specialised types of farm.	FADN	FADN: NUTS 0 and 1	1990 and 2000
			Diversification is indicated by the share of agri-environment payments in gross farm income.	FADN	FADN: NUTS 0 and 1	1990 and 2000
	17	Marginalisation	Share of holdings with low Farm Net Value Added per Annual Work Unit in combination with a high share of holdings with farmers close to retiring age.	FADN	FADN: NUTS 0 and 1	1990 and 2000

Domain/ sub-domain	No	IRENA indicator	Headline indicator and *sub-indicators*	Data sources	Spatial scale	Temporal scale
Pressures: Pollution	18	Gross nitrogen balance	Gross soil surface balance for nitrogen.	OECD website and EEA calculations on the basis of Eurostat's ZPA1 data set or Farm Structure Survey	NUTS 0	1990 and 2000
	18b	Atmospheric emissions of ammonia	This indicator shows the annual atmospheric emissions of ammonia (NH_3) in the EU-15 for 1990–2002, and the contribution that agriculture made to total atmospheric emissions of ammonia in 2002.	Officially reported 2004 national total and sectoral emissions to UNECE/EMEP (Convention on Long-Range Transboundary Atmospheric Pollution)	NUTS 0	1990–2002
	19	Emissions of methane and nitrous oxide.	Aggregated annual emissions from agriculture of methane (CH_4) and nitrous oxide (N_2O). Emissions are shown relative to 1990 baseline levels expressed as CO_2 equivalents.	Official national total, sectoral emissions, livestock and mineral fertiliser consumption data reported to UNFCCC and under the EU Monitoring Mechanism and Eionet	NUTS 0	1990–2002
	20	Pesticide soil contamination	The indicator uses a model to calculate the potential average annual content of herbicides in soils.	Calculation of the total PPP quantity present in a specific NUTS 2 region is based on Eurostat pesticide statistical data (2002) and FSS (1997, 2000)	NUTS 2 and 3	1993–1999
	21	Use of sewage sludge	Use of sewage sludge in agriculture.	Data submitted by Member States to the European Commission in the context of the requirements under the standardised reporting directive (91/692/EEC)	NUTS 0	1995–2000
Pressures: Resource depletion	22	Water abstraction	Water abstraction by agriculture is indicated by the annual water allocation rates for irrigation.	Joint OECD/Eurostat questionnaire	NUTS 0	1990–2000
			Regional water abstraction rates for agriculture have been estimated by weighting the reported national rates by the regional irrigable area.	Joint OECD/Eurostat questionnaire, FSS	NUTS 2 and 3	2000
	23	Soil erosion	Annual soil erosion risk by water.	Pesera model using CLC (Land use), GTOPO30 (Relief), MARS database (Meteorology), European Soil Database	NUTS 2 and 3	2003
	24	Land cover change	Area of the entries and exits to and from agricultural and forest/semi-natural land between 1990 and 2000.	Corine land cover	NUTS 2 and 3	1990 and 2000
			Net land cover changes for arable land and permanent crop and pasture between 1990 and 20001.	Corine land cover	NUTS 2 and 3	1990 and 2000
	25	Genetic diversity	Distribution of risk status of national livestock breeds in agriculture.	FAO's Domestic Animal Diversity Information System (DAD-IS), status July 2003.	NUTS 0	July 2003
Pressures: Benefits	26	High nature value (farmland) areas	This indicator shows the share of the Utilised Agricultural Area that is estimated to be High Nature Value farmland.	CORINE Land Cover and FADN	NUTS 0	1990
	27	Production of renewable energy (by source)	Land use devoted to energy/biomass crops, and primary energy produced from crops and by-products.	Eurostat FSS and RES; European Bio diesel Board; EurObserv'ER; Fachverband Biogas; Statistics Sweden: International Energy Agency; Faostat	NUTS 0	2003
State: Biodiversity	28	Population trends of farmland birds	Population index trends of up to 23 selected bird species that are common and characteristic of European farmland landscapes.	Pan-European Common Bird Monitoring project (RSPB/EBCC/BirdLife International)	NUTS 0	1990–2001
			Share of farmland birds with declining populations.	BirdLife, EBCC (2000): European Bird Populations — Estimates and trends. BirdLife Conservation series No 10.	NUTS 0	1990–2002

Annexes

Domain/ sub-domain	No	IRENA indicator	Headline indicator and *sub-indicators*	Data sources	Spatial scale	Temporal scale
State: Natural resources	29	Soil quality	Topsoil (0–30 cm) organic carbon content.	Soil: European Soil Database, Corine land cover, Global Historical Climatology Network — GHCN, Pedo-transfer model to calculate organic carbon content	NUTS 2 and 3	(2000)
	30.1	Nitrates in water	Annual trends in the concentrations of nitrates (mg/l N) in ground and surface water bodies.	Eurowaternet	NUTS 0	1992–2001
	30.2	Pesticides in water	Annual trends in the concentrations (µg/l) of selected pesticide compounds in ground and surface waters.	Eurowaternet

Denmark: NERI (2004); GEUS (2004); Ministry of Environment (2003)

United Kingdom: Environment Agency (2004)

Austria: UBA Vienna (2005)

Finland: FEI (2001) | NUTS 0 | 1992–2001 |
| | 31 | Ground water levels | Trends of groundwater levels. | Spanish Ministry of Environment | Case study (Spain) | (1978–1998) |
| ***State:*** Landscape | 32 | Landscape state | The diversity of agricultural landscapes across Europe is shown by analysing selected landscape parameters with strong links to agricultural land use. These parameters have been calculated for selected regional case study areas representative of different European landscapes (e.g. Montados of Portugal, field landscapes in central plateau of Spain, bocage in France, Highlands of Scotland). | CLC (patch density)

FSS (crop distribution)

LUCAS (linear elements) | NUTS 2 and 3 for case studies | CLC 1990 and 2000

FSS 1990 and 2000 |
Impact: Biodiversity	33	Impact on habitats and biodiversity	1) Share of Important Bird Areas (IBA) in the EU-15 affected by agricultural intensification and/or abandonment	IBA programme of BirdLife International	NUTS 0	2004
			2) Population trends of agriculture-related butterfly species in Prime Butterfly Areas	Survey of prime butterfly areas by Butterfly Conservation International.	NUTS 0	2003
Impact: Natural resources	34.1	Share of agriculture in GHG emissions	Contribution of the agricultural sector to total EU-15 emissions of the greenhouse gases CO_2, CH_4, and N_2O.	Official national total, sectoral emissions, livestock and mineral fertiliser consumption data reported to UNFCCC and under the EU Monitoring Mechanism and Eionet	NUTS 0	1990–2002
	34.2	Share of agriculture in nitrate contamination	Nitrogen emissions to water by economic sector.	OECD website and UBA, 2001	NUTS 0	1990 and 1998
	34.3	Share of agriculture in water use	Share of agriculture in water use from surface and ground waters.	Joint OECD/Eurostat questionnaire

FSS (variable irrigable area — area covered with irrigation infrastructure) | NUTS 0 | 1990 and 1998 |
| ***Impact:*** Landscape | 35 | Impact on landscape diversity | Trends of indices of overall agricultural diversity. This indicator presents the evolution of some of the parameters calculated in IRENA 32. The changes of the crop type distribution (e.g. arable, grasslands) and patch density are shown for the selected landscape types. | CLC (change number of agricultural classes and patch density)

FSS (change in crop areas) | NUTS 2 and 3 for case studies | 1990 and 2000 |
| | | | Changes in total linear landscape features (km). | UK Countryside survey (data for England, Wales and Scotland)

Swedish Countryside Survey — Monitoring landscape features, biodiversity, and cultural heritage (LiM project) | NUTS 0 (United Kingdom, Sweden) | 1990–1998 |